Mines and Independence

A Future for Namibia 3

Mining

Mines and Independence

The Catholic Institute for International Relations

First published in September 1983 by
Catholic Institute for International Relations, 22 Coleman Fields, London
N1 7AF.

ISBN 0 904393 77 1

Mines and Independence. (A Future for Namibia; 3)
I. Mines and mineral resources — Namibia
I. Catholic Institute for International Relations. II. Series
333.8'5'09688 TN119.N/

Copies available from CIIR and the Africa Fund, 198 Broadway, New York
NY 10038, USA.

Trade distribution to bookshops and library suppliers by Third World
Publications Ltd., 151 Stratford Road, Birmingham B11 1RD, Tel: 021-773
6572.

Printed by the Russell Press Ltd., Bertrand Russell House, Gamble Street,
Nottingham NG7 4ET.

Design by Jan Brown Designs, London.

Contents

Tables in Text

Tables in Appendix

Figures

Maps

Preface

CIIR is very grateful to the many people who provided information, comments or suggestions, and read the various drafts of this booklet. Some must remain anonymous, but in particular we acknowledge the comments of Peter Agar, Martyn Marriott, Lawrence Morris and Richard Moorsom, and the extensive help of Professors R.H. Green and Peter Walshe of CIIR's Education Committee. Consultations with SWAPO officials were also most helpful. Mining companies provided information, as did mine workers, church sources and other individuals inside Namibia. The final interpretation and analysis, however, is that of CIIR alone.

This study has benefited greatly, as must any study on mining in Namibia, from three key documents and from discussions with their authors: the two United Nations Institute for Namibia *Working Papers* on mining by H.S. Aulakh, W.W. Asombang and C.M. Ushewokunze, and Roger Murray's work *The Mineral Industry of Namibia: Perspectives for Independence* published by the Commonwealth Secretariat in 1978. CIIR is also particularly grateful to Roger Murray and the Commonwealth Secretariat for permission to look at a first draft of the unpublished revision of this original study which is expected to appear later in 1983.

1 Introduction

Mining has been at the heart of colonial Namibia. On the one hand, men leave their families behind as they go to work on the mines — this whole 'contract' system was based on the demand of the mines (and also ranches) for labour. On the other hand, the South African occupation of Namibia is supported by taxes and exports from the mines. The first two chapters explore how the mines grew up, and the position today.

The second half of the book is concerned with how the mining sector can be used in a future, independent Namibia to support a development plan that meets the needs of the people as a whole. The great advantage of mining is its ability to provide the government with funds; in other respects, we shall suggest, it presents problems.

Readers unfamiliar with the geography of Namibia may find it helpful to refer to Appendix 1, which offers a lay guide to both the geological environment and the prospects for new mines.

Map 1. Namibia

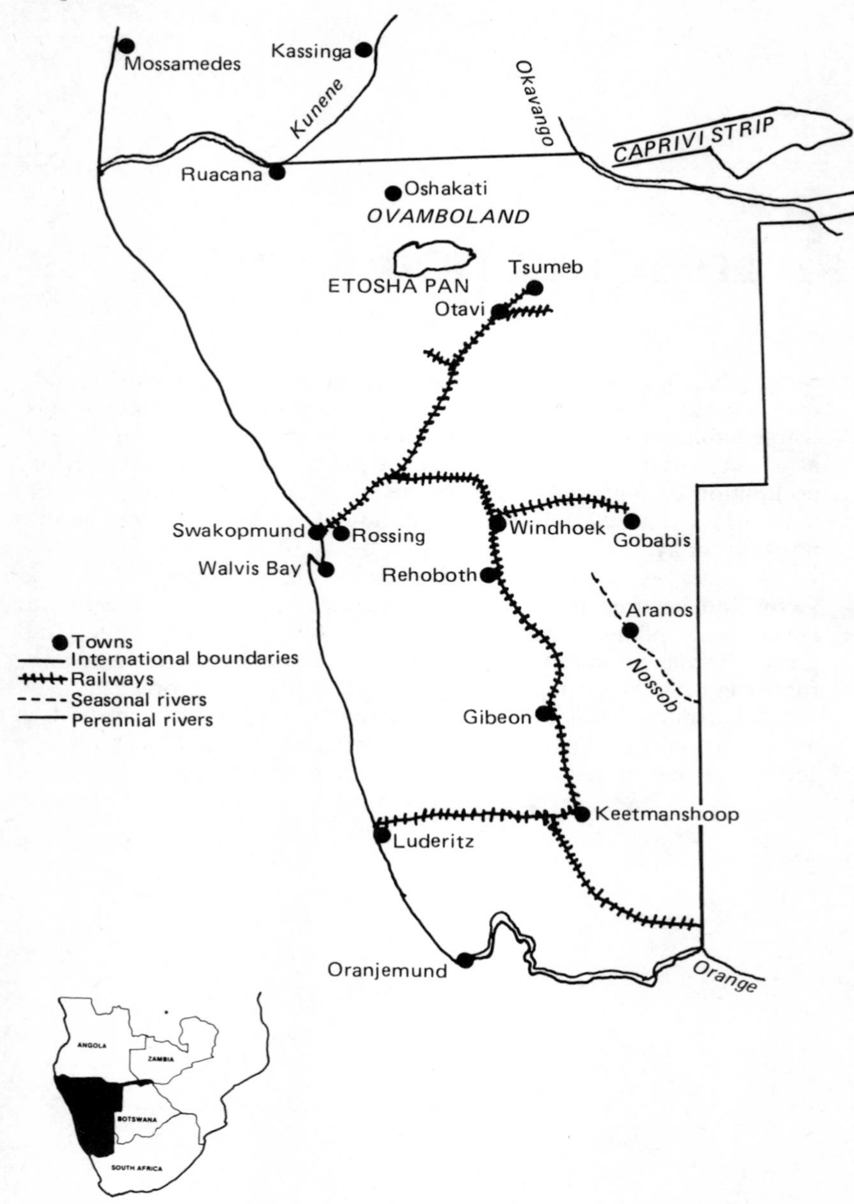

Map 2. Principal Mines

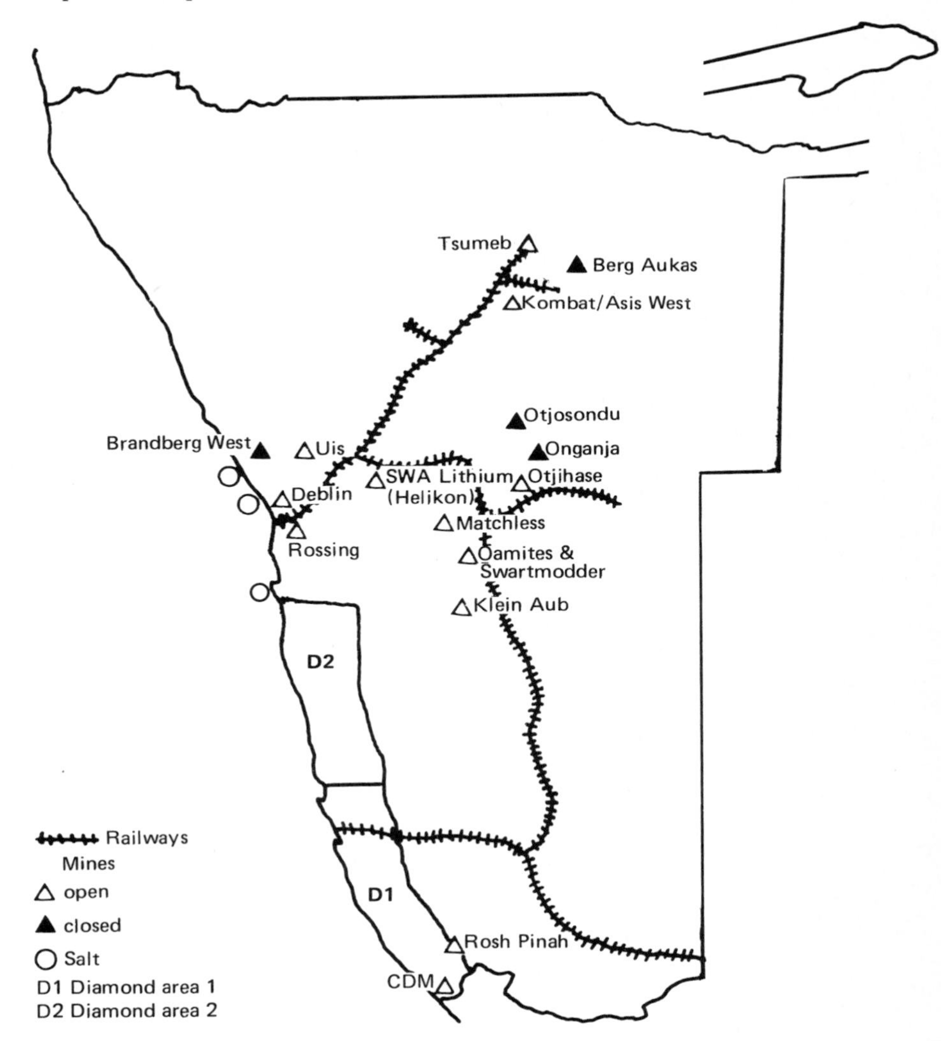

2 Mining in the Past

2.1 Pre-Colonial

Long before the Germans occupied Namibia, minerals were mined and worked: copper in the Otavi region, iron ore from Kassinga in southern Angola, salt in many places but especially to the north of Etosha pan. Up to the 19th century there was long-distance trade. Ores were moved to northern Namibia for smelting and manufacturing. Metal goods were sold to southern Angola, central and southern Namibia, and east along the Okavango river to parts of present-day Botswana. The Ovambo states of the north were clusters of dispersed homesteads, none of them larger than 80 000 people. Differences in wealth among people were small, and considerable popular control could be exercised over the kings. It is noticeable, however, that in the two clusters that grew largest, and whose kings became most powerful, the kings obtained control of the mineral trade — Ondonga over iron, and Ukwanyama over copper. Traders had to pay compulsory 'gifts'; then as now money from minerals provided a strong base for governments.

Minerals were then — as they no longer are — part of an integrated local economy. Smiths worked metals into implements or ornaments which were used in the region. Most people herded animals and, where the climate permitted, grew crops. Farmers required hoes, which were locally made.

This economy disintegrated as European traders introduced new goods (such as guns), created debt and showed an insatiable demand for ivory and cattle. Ovambo metal traders lost their southern markets by the 1890s, as mass-produced goods became available. In Ovamboland itself metal-working persisted: locally-made hoes were still in use, and bellows of the type used for metal working can still be seen, but the mines no longer met local needs.

2.2 The Coming of the Companies

> I should be pleased if it turned out that the entire soil is a colossal mineral deposit which, once it is mined, will leave the whole area one gaping hole.
> — Adolf Luderitz, German trader, 1884

Mines have a lure of high profits that sets them apart. For the successful mine, this is no illusion, and has been recognised by economists in the so-called 'theory of rent'. A manufacturer of bread, say, has to compete with other bakers who use much the same machinery and so have costs much the same as his. To earn high profits, given the bread price, will require cutting his costs hard, which is difficult and may, for example, meet worker resistance. This is much less true in mining. In some mines costs will be much lower than in others — because there is more mineral per ton of ore,[1] because the ore is nearer the surface, because transport distances are shorter, and so forth. Large profits therefore await anyone who can discover a *low-cost* mine.[2]

It is no surprise, then, that news of the existing mines, and rumours of gold and diamonds, sent European traders such as Luderitz in search of mining rights. Chiefs were induced to sign concessions, often with scant regard to whether they had such powers, or even whether they resided in the area or were acknowledged as chief by the people. The divisions among 19th century Namibians prevented a co-ordinated response, save for one remarkable moment in 1858 when most of the leaders of central and southern Namibia met at Hoachanas and signed a peace treaty. The meeting followed South Africa's first copper boom, and the opening of two small mines west of Windhoek. Article 5 of the treaty begins: 'No chief may permit copper being mined in his territory without the knowledge and agreement of all other chiefs . . . .' Today Namibians are asserting the same rights to control the country's minerals.

Colonial occupation

Namibia was allotted to Germany in the European partition of Africa at the 1884 Berlin Conference. Some ten years later, German troops undertook a military occupation, which culminated in the great revolts of 1904-06 and the deliberate genocide of much of the population of central and southern Namibia. White immigrants took over much of the land.

---

1. Ore is the mineral-bearing rock.
2. This is to over-simplify: in a world recession, even a low-cost mine cannot guarantee profits if prices plummet and there is considerable debt to be repaid.

Mining concessions passed from small prospectors to larger syndicates, financed from Germany, Britain and South Africa. Prospects, however, were not promising, and the syndicates did very little. The first government Mine Department lasted six months in 1888, during a rush after some Australian adventurers faked a gold discovery. Initial attempts at copper mining in central Namibia were abandoned in 1903. An expedition to the Tsumeb area reported in 1901 that the development of a mine there would never pay for the necessary railway. The colonial authorities despaired of the companies. Nevertheless in 1903 a start was made on the Tsumeb railway. Conditions were so bad — for the railway must be as cheap as possible — that imported Italian labourers struck and deserted in droves. The Great Revolt increased the importance of the line to the Germans and it was completed under armed guard by the forced labour of captured Herero men and women.

Tsumeb proved more profitable than orginally expected, but the real break came with the discovery of diamonds near the sea in 1908. Mining was low-cost, since little more than shovels and brooms were required by way of tools, and the large numbers of workers required came cheap. By 1913 sales ran at R4.5 million, of which only R100 000 had to be paid to the 3 000 workmen. The first world war caused an interruption; South Africa replaced Germany as the colonial power, and ownership of the diamond mines passed into South African hands and eventually to De Beers. Until the late 1920s mining continued to dominate production in the colony (see Table 1), with a number of new, smaller mines opening, particularly for tin and vanadium. A third of total national income was taken, as profits or salaries, by foreigners.

Table 1

Mining's Contribution to National Income to 1954

	£ million						
	1924	**1929**	**1934**	**1939**	**1944**	**1949**	**1954**
Total GDP	4.5	6.5	2.7	6.5	9.9	20.9	52.0
Mining	2.0	2.6	−0.03	0.7	1.1	5.5	17.3
Percentage	*44*	*40*	*0*	*11*	*11*	*26*	*33*

Source: Krogh 1960 *Note:* See Table A7 for later years

The effect on the local economy

Production was, of course, for the world market, and the effect on the local economy was very different from that of pre-colonial mining: in some ways very limited, in others very disruptive. Very little supporting industry was stimulated — most mining inputs were imported; the coke for Tsumeb came from Germany and later from Natal. Some food and timber came from the mines' own farms or to a lesser extent from settlers, both on land from which Namibians had been removed. A major impact of mines was in supplying tax revenue and bringing into being parts of the railway network. Both revenue and railways were largely devoted to bolstering unprofitable white settler farmers. These new farmers reinforced South African political control over Namibia, and were the foundation for the increasing political integration of South Africa and Namibia.

The 1930s slump

In the 1930s the world market slumped — and with it, therefore, Namibian mining. Tsumeb mine was closed between 1932 and 1937, as were the vanadium mines at Karavatu and Uris (though not elsewhere). At Consolidated Diamond Mines (CDM), diamond production was cut back at the end of the 1920s, and their mines too shut between 1932 and 1935, as a part of De Beers' overall strategy to keep diamond prices up. Minerals, which had been 42% of everything produced in Namibia in the 1920s, were only 14% in the 1930s and most of that was concentrated in the two years before the second world war. Government revenue shrank. The companies made losses, but hung on to their mines in the hope of better times to come — a practice which may be important to notice in the context of the 1980s slump. There was still money to be realised: official figures suggest that from 1932 to 1936 between a quarter and a half of GDP was sent abroad. At times the sector even repatriated more than the value of its production. But there was almost no reinvestment. As a colony, Namibia was no longer a success, except in providing land for settlers, and in denying mineral resources to others.

2.3 Labour in the Mines

> This giving of yourself into the hands of the whites will become to you a burden as if you were carrying the sun on your back.
>
> Hendrik Witbooi, 19th-century Namibian leader.

The mines required cheap labour, and yet the potential labour force in central and southern Namibia was decimated in the German genocide of 1904-06. Workers were brought in from South Africa, and plans were laid to bring indentured labour from as far as India and China. In the event, however, the colonial authorities opted for men from the

Table 2

Mine Wages (according to companies)

Minimum Wages[1]

Rand per Month

	1893	1913	1922	1946	1954	1961	1971	1973
Minimum Wage — actual	2.00	2.50	1.95	1.95	3.25	4.55	8.69	24.96
— in 1938 prices	n.a.	n.a.	1.55	1.39	1.49	1.72	2.22	5.61

Note

1. By comparison, the CDM minimum wage in early 1982 was R230 per month, and Tsumeb's R110-120 for single and R150 for married people. In early 1983, the CDM minimum was R250 per month, Rössing R296 and Tsumeb 59 cents/hour or approximately R123 a month.

Sources

Gordan 1977; price index from Odendaal 1964, *Statistical/Economical Review* 1982. Companies

Average Cash Wages for Unskilled Workers[1]

		1973	1977	Early 1982
Tsumeb[3]	— actual	37	(76)	(140)
	— 1973 prices[2]	37	(48)	(51)
CDM	— actual	87	210	398
	— 1973 prices[2]	87	134	144

Notes

1. Food is also provided. In 1973, TCL valued food at R35 per worker, and CDM at R25.
2. These are deflated by the Windhoek consumer price index.
3. 1973 figure from Gordon. Thereafter, TCL figures are based on US dollar figure for 1975 ($64, taken to be R47), from Newmont, plus the annual % increases reported in TCL Annual Reports. They should therefore be treated with extreme caution.

Sources

Gordon 1977 p. 8; CDM; *TCL Annual Reports,* Murray 1978, quoting Newmont.

north of Namibia and southern Angola, especially from Ovamboland. Unlike the rest of the country, the north remained 'indirectly ruled', under local kings rather than under direct colonial administration. Yet people, and especially the kings, required income to buy guns and Western goods and to pay debts to European traders. Farming in the area was vulnerable to drought and animal disease, and in any case the colonial government prevented most sales of animals or crops. To earn cash, men therefore left for short-term work contracts in the south, making 'gifts' to the kings on their return. After 1907 numbers increased rapidly, and by 1910 10 000 Ovambo men were taking six-month contracts in the south. In 1926 the two chief mining companies themselves set up the Northern Labour Organisation (later SWANLA), with offices in the north to organise the flow of recruits. Numbers fluctuated, depending on the state of the world economy, which determined the level of activity on the mines: the influence of the world market permeated the whole country.[3]

This migrant labour system was eventually reinforced by legislation preventing families from following their men, but even without the legislation there was little choice. Most mine workers were and are married.[4] Wages and housing on the mines were intended to provide only for single people, whose families could be assumed to keep themselves by farming.

Working conditions

Conditions on the mines were generally appalling. Table 2 shows the course of minimum wages, with little change in real terms for half a century up to the 1970s — indeed rates in 1954 were lower than in 1922. Average wages would be above these minimum figures (especially on the diamond mines), but even in 1973 the absolute bare minimum 'household subsistence level' for a family to live on was estimated at more than three times the minimum wage.[5] Wages were also lower even than on the South African mines, to the extent that people from the east of Namibia illegally slipped into Botswana in order to enlist for South Africa.[6]

Workers lived in compounds, 12 or more to a room. In 1918 the general level of health was so low that almost half the Africans — but

3. This is inevitably a schematic account. For more detail see Moorsom 1977.
4. 76% of a sample at Otjihase in 1974, 79% of a 1969 sample at Tsumeb (Gordon 1977, p. 84).
5. Gordon 1977, pp 8,11.
6. Including the Herero who had fled to Botswana from genocide in Namibia and in 1907 concluded a contract with a South African mine, at a time when Namibian companies were, ironically, very short of labour. There was also legal recruitment for South Africa in the 1940s and 1950s inside Namibia.

only 8% of whites — died in an influenza outbreak at Tsumeb. Supervisors could be brutal — in two months during 1911, 15 cases of physical maltreatment of workers on the diamond fields actually reached the courts. The 1917 Mines and Works Proclamation gave employers near-dictatorial powers, and made even absence from work a criminal offence. A battery of laws in the 1920s and 1930s confined black people to 'reserves', forced them to carry passes and controlled their movement.

Resistance

The human consequences of this migrant labour system have been deeply felt. They have been reported elsewhere, particularly by the churches. Divided families were, and still are, placed under great strain. The daily workload of the women left behind is much increased. The attitude of the workers themselves is summed up in a letter sent from workers at Walvis Bay to those in Windhoek, urging support for the 1971 strike against the contract system:

> Greetings. Love your neighbour as yourself, thus say the commandments of God. We don't want the contract anymore because we have no human rights. It denies us recognition as human beings. Due to the contract the Ovambo people are not regarded as people.
>
> Mr M.C. Botha (a South African official) said at Oshakati on 15 November that we, the Ovambo people, prefer to work under contract. Here Mr M.C. Botha declared that we ourselves want the contract. Now we don't want the contract anymore and want to be regarded as human beings. If the contract is not withdrawn, good reason must be given why it is not withdrawn.
>
> Now we claim our right to freedom of movement and to look for work wherever we want in the whole of Namibia. If the contract is not abolished, we will not come back to work. Why is it that if a Boer brought me and he doesn't want me anymore he can send me back to Ovambo, but if I don't want, I have no right to leave the work and go some other place? The word of God says Christ died to free all the people, but we Ovambo are not free of the contract. God did not place hate between person and person. If he did, give us chapter and verse where it is written in the bible, just as we gave it at the start.

The 1971 strike emerged from a history of resistance. The earliest reported strike on a mine was in 1893 for an increase in wages. At least six others are recorded up to 1950 and 22 more between 1950 and 1971 with more presumably unknown.[7]

In general, control was too tight to permit successful strikes, or

7. Gottschalk 1978.

union organisation. Nevertheless mine workers maintained a whole network of informal links and mutual support. A researcher at the Otjihase mine in the early 1970s commented: 'It is difficult to describe the intensity of the solidarity among the migrant workers of the mine.'[8] If a supervisor ill-treated a miner, the whole gang quite frequently walked off to lodge a complaint. In the difficult living conditions of overcrowded hostels, cooperation was essential and unity was forged. A united front was maintained in the face of white management, since, in the words of an Otjihase worker, they 'are at a far end of the scale, and ourselves at the other, so we could not be expected to agree on any matter'.

2.4 Post-war Success for the Companies

The end of the second world war saw the start of a long boom in the West, with minerals in demand. Trans-national mining companies reaped profits and sought out new supplies. In Namibia, the colonial extraction of wealth moved to a new level. The scene was set by the sale of the Tsumeb mine, which, until seized as enemy property in 1940, had remained owned by the German company OMEG specifically formed to exploit the mine. It was now bought, for R2m., by a consortium of United States- and British-based transnationals, headed by AMAX and Newmont. The initial outlay was recouped in the first year by working the ore dumps, and a major expansion programme was begun. Diamonds too were in high demand, so that by 1953 the diamonds stockpiled by De Beers in the 1930s had been sold off at a considerable profit, and production was expanding.

Mechanisation

Mining was on a larger scale than before, with high technology and a greater input of capital. Both Tsumeb and CDM embarked on major investment programmes in the early 1960s. Some two-thirds of mineral income came from diamonds. Diamond production, which had fallen from around 600 000 carats per year in the late 1920s to 150 000 in the late 1940s, rose to over 1.5m. in the late 1960s.

This rise was achieved by mechanisation. At CDM, giant earth-moving machines removed the covering layers of sand as well as the diamond-bearing gravel along the beaches. Processing was also increasingly mechanised. Workers' jobs changed to match — from crawling over sand dunes on all fours to controlling large machines.

---

8. Gordon 1977, p. 105.

At Tsumeb too production rose, with twice as much ore mined on average in the 20 years since the second world war as in the peak pre-war year. The main innovation, however, was to process the ore further before export — to the stage of smelting copper and refining lead, which also permitted the extraction of silver and other highly-priced by-products. The technology had to be specially developed to suit Tsumeb's complex ores, and has subsequently been used to refine ores from other countries on contract.

At other mines too, the characteristic has been for large companies to adopt technically advanced methods: Falconbridge of Canada at the Oamites copper mine in 1971, Johannesburg Consolidated Investments at Otjihase copper mine in 1975, and most recently RTZ and others in the R350m. Rössing Uranium Mine. In between the major projects, a number of small, primitive mines sprang up and withered in response to fluctutations in price: in the Rehoboth Gebiet, for example, small amounts of kyanite and zinc were mined in the 1950s and 1960s, and even gold during the second world war. But these were the exception. A wide range of transnational companies became involved — from France and West Germany as well as Britain, the United States, Canada and South Africa. The boom was not smooth, but the trend was clear.

The outflow of wealth

Mining became an enormous suction pump, extracting minerals and wealth. The minerals themselves were and are almost entirely exported: from a value of R2.5m. in 1945 they rose to R677m. in 1978 with the start of the Rössing mine. Part of the proceeds is spent abroad to pay for imported inputs to keep the mines running. Another part ends up as profit, which is foreign-owned and therefore also often sent out of the çountry. An official South African commission reported that between 1943 and 1962 CDM made R369m. profit, of which two-thirds was paid out either as dividends (R138m.) or tax (R105m); Tsumeb between 1946 and 1961 made R140m., and paid 90% of it as dividends (R91.5m.) or tax.[9] White employees also send out some of their earnings. For the economy as a whole — of which mining was by far the leading sector — by 1977 over a third of the income generated left the country, according to a UNIN estimate.[10]

9. SA, Odendaal Commission 1964, pp 331ff. On the other hand, SWACO was struggling, paying no dividends between 1955 and 1962.
10. The South African estimate was 17½% of NDP (*Statistical Economic Review* 1982, p. 29).

Figure 1

Who Benefits from Mining: 1974-78

Breakdown of total sales income

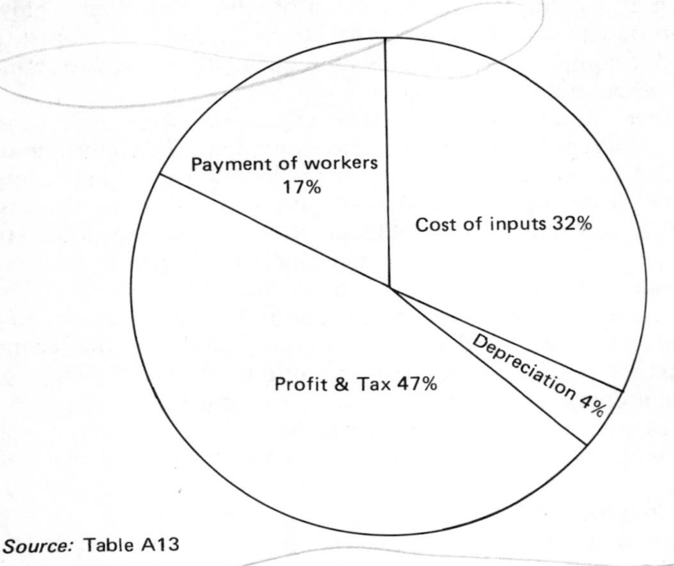

Source: Table A13

What then did Namibians gain to counterbalance this great outflow? Fig.1 and Table A13 show an overall picture for mining, from South African figures. In the mid-1970s around a third of the income from sales was paid out for inputs, transport and services of one kind or another. A full half of the income was taken as profit. Of course, some of this profit was taxed, but Namibians had no say in how the tax was spent. That left less than 20% of sales income for wages and salaries — but the greater part of that was paid to white workers. *Perhaps 5% of the value of mineral sales ended directly in the hands of black Namibians in 1977.*

2.5 Labour Conditions Since the 1940s

Until the 1970s, workers benefited little from the post-war boom, though conditions at CDM were better than elsewhere. Wages rose very slowly in real terms (Table 2). A reporter from the British *Guardian* newspaper in 1973 described conditions in the labour compound at Berg Aukas mine as bleak: 'There are no recreational facilities; the kitchens are poor and, when I visited them, dirty and swarming with flies.'[11] Above all, the 'contract' system of migrant labour remained intact.

Opposition and the 1971 strike

Meanwhile opposition to 'contract' was rising. The Ovamboland People's Organisation, which grew into the liberation movement SWAPO, was founded in 1957 explicitly to fight the contract system, and it was in the contract workers' compounds that its support was rooted and its ideas transmitted. As a CDM official put it, no better way than compounds of assisting SWAPO to recruit could easily be imagined. 'Contract' brought together Namibians from all over the country, creating both a sense of national unity and, as migrants moved to and fro, a national communications system. Compounds gave SWAPO a place to educate and organise.

In December 1971 a massive strike broke out, when almost half the Ovambo migrant workers returned home to prove their opposition to contract. Workers struck at Tsumeb, Berg Aukas and nine other mines. At CDM an initial stoppage ended with management agreeing to relay workers' concerns to the government, and shortly afterwards wages were raised 10%. For months workers remained in Ovamboland, where the South African army was deployed, a state of emergency declared and repression unleashed.

In the middle of the strike, the South Africans agreed to abolish the SWANLA labour recruiting agency. A new contract called an 'agreement' was imposed, which supposedly made it easier for workers to change jobs. In the event, these changes proved more cosmetic than real — labour bureaux run by bantustan 'authorities' replaced SWANLA, and it was still impossible for families to move to the place of work: the men went alone. In the words of one worker replying to a questionnaire sent out through church channels after the strike: 'It is said that one has the right to work freely, that one works together with one's master. In reality it is like cat and dog.'[12]

The churches and the mines

The majority of Namibians are Christian, and so therefore are most mine workers.[13] Their opposition to exploitation has been informed by their Christianity, as shown by the 1971 strike letter quoted earlier. The major churches themselves were, until around 1960, dominated

11. Adam Raphael in *Guardian* 8 May 1973.
12. Quoted in Green and Kiljunen, 1981, p 129.
13. 92% at Otjihase in 1973 (Gordon 1977, p 84).

by foreign missionaries. On the whole, the churches saw their role as to provide Namibians with religious services narrowly conceived and to perform works of charity. After the withdrawal of foreign mission control in the Lutheran churches, however, the new Namibian church leadership came to challenge the legitimacy of the South African occupation. At the same time, the churches came increasingly to oppose the contract labour system.[14] During the unrest before the 1971 strike, Bishop Auala of the Lutheran ELOC Church was called by the Chief Native Affairs Commissioner to a mass meeting; after hearing the workers, he answered, 'You have no choice but to go on strike.' During the strike itself six Anglicans were shot dead by the South Africans whilst leaving church in Ovamboland; Bishop Winter was deported from Namibia after conducting an enquiry. When the National Union of Namibian Workers began in 1978, one of the organisers was Pastor Gerson Max, the ELOC minister for migrant labourers.

Changes in the 1970s on the larger mines

The 1971-72 strike had an influence on labour conditions. Average cash wages at Tsumeb rose 60% between 1971 and 1973. There was another large rise, of 51%, in 1976, but since then wage rises have scarcely kept up with inflation and in absolute terms Tsumeb wages are still very poor. In 1982, the minimum wage for a married man was said to be around R150 per month, and for a single man between R110 and R120; by contrast, in March the same year it was estimated that a family living in Windhoek required a minimum 'Household Subsistence Level' of R261 to live on.

A major change occurred at CDM, whose wages in the 1960s had been much the same as Tsumeb's. By 1973 its average cash wages were more than twice Tsumeb's, and in early 1982 the minimum wage was R230.[15] Rössing tended to follow much the same path — in early 1982, the average wage at both mines was said to be R398 per month.

These rises in wages on the big mines were accompanied by increased spending on houses. In the smaller mines no more than a handful of rooms were set aside for wives to share with their miner husbands on brief visits. Only Rössing proposed to provide family housing as the general rule — and even so a quarter of the African

---

14. Both a challenge to South Africa and an attack on contract labour were contained in the 1971 Open Letter from the two main Lutheran Churches to the South African Prime Minister.
15. These figures are not strictly comparable between mines, as there are different deductions for rent etc, and provision of meals may vary. The Tsumeb minimum is said to be linked to UNISA and University of Port Elizabeth estimates of the 'Household Subsistence Level'.

workers were still in a compound ('Rössing Village') in 1983. CDM and Tsumeb built family housing for their middle-level African staff, some 5 to 10% of the total. Alongside went a policy of training a limited number of blacks to fill semi-skilled and some skilled positions (see Section 4.1.2 below). The mining companies put pressure on the South Africans to repeal the law preventing blacks from taking skilled work.

Reasons for the change

This change in the behaviour of the transnational mining companies towards their workers came about for three reasons. Most immediately, there was a severe shortage of white artisans, all the more as the intensifying war in Namibia conscripted some whites and discouraged others from staying. The shortage also meant that the white artisans commanded ever higher wages. Secondly, the 1971 strike, the organisation of the National Union of Namibian Workers (NUNW) on the mines in 1978-80, and the imminent possibility of independence put pressure on the companies. The companies had a considerable interest in trying to satisfy the workforce, and to create a group of senior black workers more committed to the status quo on the mines. Thirdly, the advanced technology of modern mining allowed higher wages to be paid: the wage bill was only a small proportion of total costs, so that quite large wage rises could have quite small effects on the total cost of mining (Table 4).[16] Adverse international publicity may also have affected the transnational companies.

These changes have not altered the fundamental criticisms made by miners. The large mines introduced non-racial grading systems in the 1970s, with pay based on skill levels, but, since in general whites are more skilled, the differences between the pay received by whites and by Africans remain.[17] Whites retain almost all supervisory posts. Most workers are housed without their families. On the smallest mines, conditions are as appalling as ever. In April 1982 the Newsletter of the Council of Churches in Namibia reported on a visit to the Deblin mine, near Swakopmund, which employs around 86 workers:

---

16. For a generalisation of this argument, and a comparison with other sectors (such as ranching) which cannot afford such rises, see Moorsom 1980.
17. At Oamites mine in 1980, the ratio was six to one (ENOK Rehoboth Report); three to one for the median workers at Rössing in 1982, ignoring overtime and the better housing and fringe benefits for whites. In the economy as a whole in 1979, the average white income was estimated to be twenty times greater than the average black income of R160 p.a. (R. Green 1981).

Average salaries per month was around the area of R50.00. No provision is made for housing and labourers have to build own 'homes' from oil-drums from the mine and any piece of material that can give them shelter. As I entered the 'slum', stench filled my nostrils — an element of decay at a mine which is understood to be profitable. There are no lavatories and people relieve themselves behind the shack and behind a small hill 50 metres away.

Throughout the mines, church sources report that miners see the companies as part and parcel of the colonial occupation.

2.6 Namibia's Part in Supporting South Africa

South Africa's control of diamonds: building a transnational corporation

South Africa's benefit from Namibian mining stretches back to its initial occupation of the territory. As early as 1919 Sir Ernest Oppenheimer's acquisition of the previously German diamond mines of Namibia was crucial to his establishing of control over the world diamond market. With Namibia's CDM behind him, he was able to take over De Beers. The Anglo-American/De Beers group which he formed was and is at the centre of the South African economy.

In the longer term De Beers' principal strength has been the ability to regulate world diamond sales to keep up the price, and minimise fluctuations. Here again Namibia played a vital role. CDM has been used as one of the main regulators or taps controlling the flow of diamonds on to the world market. Production has been stepped up when demand was high — in 1962, for example, and 1969. Conversely, when the diamond market was weak, CDM mining has been cut back — for example, by a third between 1927 and 1928, or par excellence when CDM was shut down in the 1930s. The pattern has been repeated in the 1980s crisis.

Namibia as a colony

In the boom years after 1945, Namibia became a classic colony. Mineral exports earned a steady stream of foreign exchange for the South African Reserve Bank — equivalent to almost a quarter of South Africa's own non-gold mineral exports in the 1960s.[18] On the other hand, for most of the time Namibia did not cost South Africa anything: until the late 1960s, tax revenue, largely from mines, usually

18. 9% if gold is included.

Table 3

Importance of Selected Namibian Minerals to South Africa

	Unit	1977	1978	1979	1980	1981
Lead						
SA Consumption	000 short tons	40.5	43.7	48.3	58.0	54.0
SA Production		0	0	0	93.8	108.6
Namibian Production		45.4	42.5	45.2	52.6	39.7
Tin						
SA Consumption	000 tonnes	2.2	1.9	2.0		
SA Production		2.9	2.9	2.7		
Namibian Production		1.0	1.0	1.0		
Zinc						
SA Consumption	000 tonnes	56.7	71.7	78.2		
SA Production		72.4	68.7	56.5		
Namibian Production		38.3	36.6	29.0		

Notes

1. In 1971 South Africa had no zinc production and in 1974 only 32.8 thousand tonnes.
2. Production figures are for metal content in concentrates.

Sources: American Bulletin of Metal Statistics; Metal Statistics; World Mineral Statistics.

paid for the costs of running the colony.[19]

Namibia became a place in which South Africa carried out economic activities which it was not possible or not so lucrative to carry out at home. It was a source of minerals that South Africa did not possess (Table 3). The state-owned Iron and Steel Corporation ISCOR, which was a driving force of South African industrialisation, opened the Uis tin mine in 1958 and expanded it in the 1970s: in 1981 Uis was still producing half ISCOR's tin requirements. ISCOR's reasons for opening the Rosh Pinah lead/zinc mine are equally clear: until 1980 South Africa itself had almost no lead production at all, and Rosh Pinah was also supplying most of ISCOR's zinc for galvanising. South Africa's Industrial Development Corporation (IDC) was important in bringing the Rössing and Oamites mines into production.

Namibia was also a back yard for South Africa's private sector. Several Anglo group companies have mined or prospected.[20] The next largest is the Afrikaner-based mining group Federale Mynbou/Gencor, who operate the Klein Aub copper mine, hold the Langer-Heinrich uranium prospect, and who explored heavily for other mines in the early 1970s. Finally, Namibia's mines, like the rest of the economy, became a protected market for South African suppliers. Sheltered from international competition behind South Africa's high tariff barriers, they supplied mining inputs, food and services.

Namibia reached the mid-1970s, therefore, with a technically-advanced mining sector, oriented to the outside world, relying on workers separated from their families, and operating for the benefit of foreigners and specifically of South Africa. Revenue flowed to white settler farmers, to the neglect of the many small farmers in the bantustans. The contrast with pre-colonial mining could hardly have been greater.

---

19. For a decade from the late 1960s South Africa provided funds for a large infrastructure investment programme, particularly in water and electricity supplies for mines and towns. More recently, the rising costs of military occupation have coincided with a collapse of revenue; South Africa has therefore had to pay more than it received.
20. Including Gold Fields Namibia (formerly Kiln Products and owners of SWACO), Johannesburg Consolidated Investments, as well as De Beers, Charter Consolidated and Goldfields of South Africa.

3 The Mines Today

3.1 The Importance of Mining in Namibia

The 1970s saw a mineral boom in Namibia. In 1970, mineral sales were R120m., of which approximately half was diamonds, a quarter copper, with the rest made up principally of lead, zinc, tin, vanadium, tungsten, lithium, silver, germanium, salt, arsenic, pyrites and cadmium.[1] The value of mineral sales increased six fold in the following decade, as a result of a rapid rise in diamond prices and the opening of the largest uranium mine in the world (Fig.2 and Tables A2 and A3). The first part of this chapter describes the sector in the 1970s heyday, and the last part, section 3.5, shows how it has been disrupted by crisis.

Minerals in the economy

Mines are at the heart of the present economic structure of Namibia. They are far more important than any other sector in terms of the money they earn. By the late 1970s, minerals were almost half the value of everything produced in the country, over four times more than all agricultural products. Minerals constituted 85% or more of goods exported, so allowing the country to be very highly dependent

1. The most comprehensive account of Namibia's mining sector is Murray (1978 and forthcoming). See also the Namibia sections of *Mining Annual Review*, and USBM *Minerals Yearbook*. In addition to the minerals listed in the text, there were reports of manganese and iron ore production in 1966, as well as graphite and beryl ore. Bismuth, tantalite, columbite, marble, wollastonite, sillimanite, fluorspar, kyanite, feldspar, gypsum and gold have also been mined in small quantities in the past, with the first seven of these (plus graphite) thought still to be produced in Namibia (*World Mineral Statistics 1976-80*, Murray).

on imports (including food).[2] In contrast to the situation in most other countries, few of the goods used by Namibians are made in Namibia: instead, minerals are sold abroad, and foreign goods bought with the proceeds.[3]

Figure 2

The Growing Importance of Mining in the 1970s

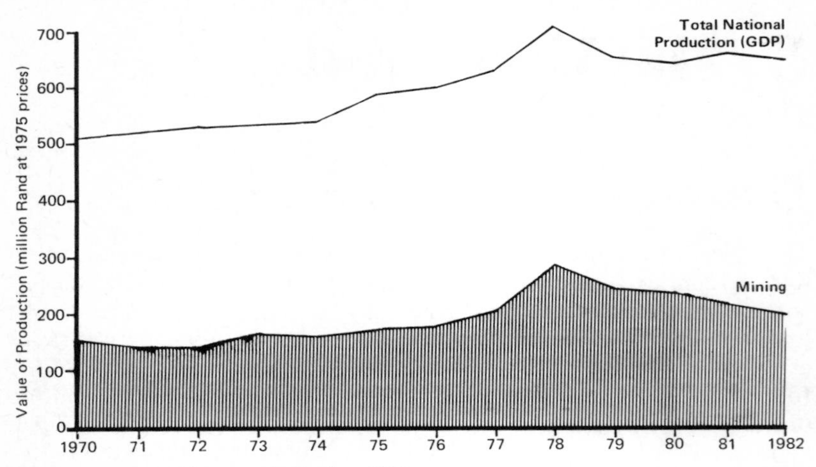

Source: Statistical/Economic Review 1982

Mining's support for South Africa's occupation

Taxes on minerals are an essential base for the present government of Namibia. Before the 1980s crisis, mines were providing directly at least one third and sometimes one half of the government's revenue (Table A8), and more if indirect taxes such as custom duties on imported mine inputs are included. Almost all the mineral tax in recent years has come from diamonds. Under the present extremely

---

2. R. Green (1981, p. 46) gives 1979 mineral exports as R850m. out of total visible exports of R1 010m. Official figures for mineral exports are not available, but almost all mineral sales are presumably exports. The share of minerals in exports would then be:

	Rm			
	1977	1978	1979	1980
Mineral Sales	581	677	774	870
Total Merchandise Exports	694	824	949	1023

Sources: Table 7 and Statistical/Economic Review 1982 (p. 36).

3. Exports were 71% of GDP in 1979, according to Green. Official figures give 80% in 1979 and 77% in 1980 (*Statistical/Economic Review* p. 36 and p. 21).

generous tax law, Rössing paid no tax until 1983 when its profits had repaid the full cost of sinking the mine. In 1981 Tsumeb Corporation was even given a tax refund because it bought Otjihase mine from its previous owners, and for tax purposes the law allows Tsumeb to deduct the capital expenditure from profits and so show a loss.

But tax is not the only way the mines support the South African occupation. Namibia is unique in the world. In no other country has the ruling regime been clearly condemned as illegal by the International Court of Justice. In ordinary national law, if somebody illegally occupies a house, bailiffs are sent to evict him. But in Namibia instead of the bailiffs the international mining companies have come. These are large and eminently respectable bodies from the principal Western countries, and number ex-Cabinet ministers among their directors. At a practical level, co-operation between company and state is continuous and inevitable, to the extent of maintaining armed units on mines which are in radio contact with the SA Police. No matter how liberal companies may strive to be, and however much they wish to be apolitical, their presence gives respectability to the South African occupation, and compromises the position of Western governments.[4]

3.2 The Importance of Namibia's Minerals in the World

How important are Namibia's minerals to the West? Namibia is a large mineral producer, according to one estimate the fourth largest in Africa and 17th in the world. Nevertheless, it would be wrong to think of these minerals as in any way 'essential to the strategic needs of the West'. On one hand it is true that Namibia's diamonds are an important strut in the delicate framework of the diamond market, but on the other hand, it is difficult to regard diamonds as vital to western civilisation. Although Namibia is one of the largest African producers of several metals, its output is a small proportion of total western production — ½% of copper, tin and silver, and 1% of lead and zinc.

Neither are there rare strategic minerals for which Namibia is a vital source. Tsumeb's cadmium is less than 1% of world supply. Lithium production is also small.[5] Namibia does have significant

4. In comparison, during Ian Smith's rebellion in Zimbabwe, international companies were forbidden to operate there.
5. Production is thought to be around 3 000 tonnes of lithium materials per year, compared with 21 000 in Zimbabwe. Lithium is used in batteries, lubricants and aluminium processing. Demand is expected to triple by the year 2000, but the USA and Chile have large, cheap resources. (*Mineral Facts and Problems,* US Bureau of Mines, 1980).

deposits of rubidium, but no major uses have yet been found for it. The only doubt attaches to germanium. Germanium is used both in electronics and in infra-red and fibre optics, and US demand is expected to more than double between 1980 and 2000. Tsumeb was in 1930 the first commercial producer of germanium in the world, and according to the US Bureau of Mines, was a major source of germanium ores until the mid-1970s. In 1980 the Bureau said the germanium-rich areas of the Tsumeb mine had been mined out, but another source has reported Namibia producing 10% of world supply and still having 17½% of world reserves in the mid-1970s.[6] There is little doubt, however, that sufficient germanium is available from elsewhere.

Namibia is an important producer of uranium. It is the fourth largest uranium producer outside the Eastern bloc countries. Its low cost 'reasonably assured reserves' are the eighth largest in the world (Table 4). Nevertheless, Namibia's position is not as strategic as this

Table 4

World Uranium Position

	Thousand tonnes of uranium			
Country	**Production 1980**	**Planned Production 1990[1]**	**Demand 2000[1]**	**Low Cost known Resources[2]**
USA	16.8	21.8		362
Canada	7.2	10.5		230
South Africa	6.1	7.6		247
Niger	4.1	12.0		160
Namibia	4.0	4.2		119
France	2.6	4.1		59
Australia	1.6	4.7		294
Brazil	0	n.a.		119
Others	1.6	5.3		157
Total	**44.0**	**70.2**		**1 747**
Western Demand	30	53-65	80-130	

Notes

1. Some of these plans have been amended — for example planned new mines in Niger have been delayed.
2. Reasonably Assured Resources recoverable with currently proven technology from known deposits at a cost of less than $80/kgU ($30/lb of uranium oxide).

Source: Uranium Resources, Production and Demand, OECD 1982.

6. USBM, 1980; P. Crowson 1977.

makes it seem, because there is plenty of uranium available: the world's mines are producing much more than is being used (p.63 below).

However, Namibia is a potential source of minerals now bought from South Africa. In the event of either sanctions against South Africa, or disruption of South African supplies, an independent Namibia would be a valuable alternative.

Importance of Namibia's minerals to South Africa
In the early 1970s, South Africa relied on Namibia for both lead and zinc supplies, but the opening of new mines within South Africa has ended this dependence (Table 3), although Namibian supplies still used. Namibia appears to be South Africa's principal source only for industrial grades of salt.

3.3 The Principal Mines

Three mining companies dominate the scene: CDM (formerly Consolidated Diamond Mines), Rössing Uranium, and the Tsumeb Corporation (TCL). Owned by huge mining multinationals, they employ 15 000 of the 18 000 Namibian mineworkers, and account for the bulk of production. We look first at them, and then at the smaller mines. (Table A1 for summary).

Consolidated Diamond Mines

Ownership, profits and production
CDM is fully owned by De Beers, itself at the centre of the Anglo-American Corporation (AAC) transnational which dominates the South African economy. The CDM mine in the extreme south of the country is essentially an earth-moving operation. Enormous tonnages of sand and gravel are moved from beaches in four locations along 100 kilometres of sea shore, and then the diamond-bearing ore is processed to remove the gems. The operation is directed from the company town of Oranjemund. Production of diamonds, which has been expanded in the mid-1960s from around 1 million carats, peaked at 2m. in 1978 and has been cut back since, in response to the poor world market.

If the mine and town had to be built again new[7], CDM estimated

7. Compare the new Jwaneng mine in Botswana, which cost around R300m.

in 1981 that it would cost R543m. The capital value of the existing plant is of course much smaller, because it is far from new. Nevertheless CDM continues to spend money on the plant. In the five years up to 1981 CDM invested R148m., although almost all of this was simply to maintain current output rather than increase production. The returns are large: in the same period, profits totalled R1240m. before tax and R560m. after it (Appendix Table A4).

The Importance of CDM to De Beers

CDM is extremely important to De Beers. In the early 1970s Oranjemund produced above 30% of De Beers worldwide after-tax profits.[8] Since then, the development of new De Beers group mines in Botswana and South Africa has reduced its share.

Nevertheless, in 1980 Namibia still provided 17% of De Beers profits, from only 10½% of the carats mined by De Beers.[9] The reason is that Namibia's diamonds are almost all of gem grade, rather than the lower grade stones used for industrial purposes, and gems are worth more per carat than industrial diamonds.[10] These Namibian gems have been a crucial part of De Beers strategy to maintain control of the world diamond market, and it is on this monopoly that the price of diamonds rests. (For further details, see below, pp98-99).

Marketing

There is close liaison in diamond marketing between De Beers and the South African government, through the Diamond Producers' Association (DPA).[11] Diamonds from Namibia are sold by CDM to another AAC/De Beers subsidiary, the Diamond Purchasing and Trading Company (Purtra), and sorted and valued in South Africa. The gems are then sold to the Diamond Trading Company (Ditra), yet another Anglo/De Beers company, who arrange for their sale to diamond merchants at its London office. Together, these companies have become known as the Central Selling Organisation (CSO).

---

8. CDM's total contribution was greater (in 1974 R81m. out of R201m.) — but part of it due to CDM's investments in other parts of the Anglo empire.
9. De Beers Annual Reports. Diamonds are measured by weighing them, and one carat equals about 0.2 grams. In 1982, CDM produced 12% of De Beers' profits, from 6% of its carats produced.
10. Over $200 per carat in 1977 compared with $7 per carat for Zaire, where most of the diamonds are industrial.
11. Diamond marketing is both complex and secretive. Timothy Green (1981) gives a useful layman's summary. The DPA consists of the South African government, in three guises (as producer through the State Alluvial Diggings in South Africa, as controller of Namibia, and as government). De Beers also operates under three hats: as De Beers, as CDM and as the Diamond Corporation (Dicor), the subsidiary which buys diamonds from outside South Africa.

Labour at CDM

CDM employed 5 470 workers in mid-1982.[12] Almost all the African labour force are workers from the north, on short-term contracts averaging about eight months. They live in hostels, 12 men to a room in the old buildings, and three in a room in the new. The company builds family houses only for skilled workers, few of whom are black.[13] Wages are the highest in Namibia with the unskilled minimum wage at R230 per month up to June 1982 (R260 thereafter), and the average unskilled earnings at R398, including overtime.

Wage differentials — what each grade gets — are based on the Patterson salary scale, as used by multinationals elsewhere in the world. This is in principle non-racial, but is designed to ensure that wage-payments reflect the existing structure of power: the more decision-making a person does, the more he is paid. As noted in Chapter 2 and in more detail later, training programmes were introduced in the 1970s. These have not satisfied the miners as they still feel black workers are only trained for one job ('once a driver, always a driver'); that workers with secondary school qualifications are sometimes left on menial jobs; and that promotion is held back.

Given the extent of repression in Namibia, CDM's workers are a surprisingly united body. They elect their own hostel and works committees, believed to be closely linked with SWAPO. Organisation is such that since 1971 the mine has three times been brought to a standstill by strike action, most recently in October 1982 when two black workers were sacked after a fight with a white supervisor. In each of the three strikes, management made concessions quickly.

Future life of the mine

The future for diamonds in Namibia depends in the first instance on whether there are more diamonds to be mined. The number of diamonds won from each ton of ore has certainly fallen since the 1960s. CDM's official position is that 'without taking into account the possibility of new discoveries as a result of prospecting, and depending on the buoyancy of diamond markets in years to come, it is

---

12. The number of workers employed by CDM is ambiguous. The figure quoted was provided by CDM in May 1982; the De Beers Annual Report, on the other hand, gives the CDM workforce as 6 541 in 1982. The clue is presumably provided in CDM 1982, which stated that CDM had 5 800 employees on the mine at any given time. The larger figure presumably includes migrant workers at home betwen contracts. Employment presumbably fell from 5 800 early in 1982, when CDM 1982 was published, to 5 470 in May; a visitor in September was given a figure of 5 160.
13. In 1982 CDM had 923 houses and 147 flats. Migrants made up about two-thirds of the resident workforce on the mine.

hoped that the life of the mine could extend to the end of the century'.[14] In practice, mining companies rarely prove up reserves for more than 15 years' production, so it is likely that, after further prospecting, the mine would last longer. R18m. was spent by CDM on diamond prospecting in 1981, along the coast, upstream on the Orange River (said to be 'interesting' and 'encouraging') and also at sea; reconnaissance prospecting over the rest of Namibia was described as 'disappointing'.[15]

CDM's View of the Future

CDM has been seeking to reinforce its own position in the event of independence in Namibia. Its headquarters were moved from South Africa to Windhoek in 1977, with a prestigious office block built on the main street. At the same time R25m. was reserved for non-diamond investments in Namibia. It is unclear how much has actually been spent, but R5m. was used for a mineral survey 'in lesser known parts of the country which are not considered prime prospecting areas', with the results made available through the state Geological Survey.

Through the Chairman's Fund, CDM spent R3m. between 1977 and 1981, and established itself as a benefactor of small projects. Above all, the company has concentrated on the higher levels of education, presumably in the belief that a group of skilled black people constitutes the best hope for stability in the future. A technical institute was built in Ovamboland to train potential apprentices.[16] R5m. has been spent on Concordia College, a state-run non-racial high school due to open in 1983, although the initial applications for entry suggest students will, in practice, be almost entirely black.

An incident in December 1982 illustrates CDM's attitude to SWAPO. Over 200 SWAPO pamphlets were removed from a worker by police at the CDM entry gate, and the local SWAPO branch wrote to CDM to complain. The General Manager wrote back to SWAPO, denying that CDM had reported the matter to police: a policeman 'happened to have been there'. No action was taken against the worker by the mine. The incident is instructive for two reasons: on the one hand, CDM is showing that it does not wish to offend SWAPO, but on the other hand, CDM remains inevitably under the authority of the present regime.

---

14. CDM (1982). The grade of ore mined fell from an average of 16.9 carats per 100 tonnes in 1965-1967 to 9.8 in 1980-1982 (Annual Reports of CDM and De Beers).
15. De Beers 1981 and 1982 Annual Reports.
16. This Valombola Technical Institute has however had considerable problems, with only 66 students enrolled for 1983. De Beers Annual Report 1982, WA 20 June 1983.

Rössing Uranium

Ownership, profits and production

In 1971 the International Court of Justice confirmed that South Africa's occupation of Namibia was illegal. Two years later some of the great mining transnationals began major capital expenditure on the largest uranium mine in the world. By 1980 its reported profits approached those of CDM's after tax (Appendix Table A5). By 1983 the companies had recovered their investment, totalling $380m. in 1980 (of which $150m. was loan).

Rössing is part of the British-based Rio Tinto Zinc (RTZ) group, who own 46½% of the equity shares, but only 26½% of the voting rights. Other shareholders include RTZ's Canadian subsidiary Rio Algom (10%), Total of France (10%), Gencor of South Africa (7%) and others including Urangesellschaft of West Germany. The South African state-owned Industrial Development Corporation (IDC) has both a 13% shareholding, and loans of at least R60m. Its voting rights are belived to be considerably higher, perhaps a majority vote giving an effective veto right over decisions; also through legislation the South African government can demand as much of the Rössing uranium as it wishes to buy.

The mine itself is a single open pit, mined by enormous electric shovels. From the pit large lorries deliver the mined rock either to waste dumps or into the processing plant. The uranium oxide is then extracted and packed into standard drums.

1982 saw a large increase in profits, mainly as a result of a 38% rise in turnover. This increase in turnover was achieved despite a slight fall in the volume of uranium produced. Perhaps at most 20% of the increase could be attributed to changes in the exchange rate between the dollar, used in sales contracts, the rand, in which Rössing operates, and the pound, in which RTZ reports. The rest of the turnover increase has therefore to be explained either by higher prices, against the trend in the world market, or by an increase in sales of uranium over 1981 even though production was down.

The Importance of Rössing to RTZ

Rössing is an important investment for RTZ. In 1982 it was only 4½% of the RTZ group's assets and only 5½% of the group's sales, but it contributed 26% of RTZ's profit after tax. Rössing is RTZ's major uranium mine, with production a third greater than Rio Algom's North American mines. In 1982, Rössing's pre-tax profit was 40% of its assets; for the rest of RTZ's uranium business, assets one and a half times larger than Rössing's yielded only 13%.

Markets

Rössing's uranium is sold on long-term contracts, primarily, it seems to Western Europe, Japan and probably Taiwan. Intermediate processing may also be carried out in the United States, Canada and possibly elsewhere (Table 5). A major contract which enabled the mine to open was with the British Government's British Nuclear Fuels Ltd, now apparently transferred to the Central Electricity Generating Board (CEGB). The history of this contract has been told elsewhere.[17] In brief, the Labour minister responsible, Tony Benn, says it was approved in the mistaken belief that RTZ would supply from Canada rather than Namibia. The Labour Party then promised to cancel it, but reversed their position after the election, (i) in the belief that uranium was not available from elsewhere, and (ii), arguing that the contract was not illegal because South Africa was in practice the administering authority in the territory. Both assertions have been vigorously challenged by SWAPO and others.

We shall see below (p.63) that the West does have alternatives to buying from Namibia. One reason for Western countries wishing to buy from Rössing is the advantage of having supplies from a variety of sources, although it is admittedly difficult to see, for example, Canada denying uranium to the West. A further reason for some customers may be that Canada, Australia and some other producers refuse to allow their uranium to be used for military purposes, and impose strict reporting procedures. South Africa is much less stringent about uranium from Rössing, and as a result Rössing's uranium has been reported to sell at a premium of $5 to $8 a pound.[18]

Like most other uranium sales, the contracts set a long term price, with provisions for raising it as costs rise.[19] By comparison, the short term, 'spot', price for uranium has swung widely. In 1976 the spot price rose well above the Rössing price, and Rössing re-negotiated a higher price with Britain.[20] Thereafter, however, the spot price has fallen sharply because world demand for uranium is much

---

17. CANUC 1980, SWAPO 1982. Other information from RTZ Annual Reports, *Fact Sheets* no.2 issued at RTZ Annual General Meetings, *Rössing* 1980, Jepson 1977.
18. WA 11 Apr. 1980, quoting *Metal Bulletin*. Niger, the fourth largest producer in the world, about equal with Namibia, also follows the same policy.
19. The sales contracts are also fixed in US dollars. When the SA Rand fell against the dollar in 1982, Rössing therefore had extra profits, because its income was in dollars but most of its costs in Rand.
20. From less than $10 a lb (1977 delivery) to around $13/lb. This was still less than for example the $20-$25 price Britain agreed to pay for South African uranium. (*Nucleonics Week* 9 Sept. 1976, 21 Oct. 1976, 23 Dec. 1976, quoted in Radetzki 1981).

Table 5

Known and Possible Rössing Contracts

Customer	Country	Country Processing	Amount (tons)	When the contract expires
British Nuclear Fuels Ltd (CEGB)	UK	UK	7 500[1]	1984 Not to be renewed
Comurhex	France	France	11 000	late 1980s
Veba	West Germany	Netherlands		?
Urangesellschaft	West Germany	Netherlands	6 140	?
Kansai Electric Power Corporation	Japan	? USA, Canada	8 200	1986
Other Japanese[2]	Japan	? USA, Canada	23 000	?
Taiwan Power Corporation[3]	Taiwan	Netherlands	4 000	?2005

Notes

1. RTZ in 1981 stated that additional amounts would be supplied.
2. Kansai and two other Japanese firms have contracts for around 23 000 tons of uranium from South Africa, including from Rio Tinto South Africa. It is assumed by MIT these are in fact supplied from Rössing.
3. Ruurd Huisman and David de Beer reported this contract with RTZ, starting in 1990. They argue that Rössing must be the prospective source, although the RTZ group does also have uranium in the USA.

Sources

Martin Bailey in *New Statesman* 28 Jan. 1983, quoting Massachusetts Institute of Technology study for US Government.
Huisman and de Beer in WO 29 Jan. 1983.

less than anticipated. The price Rössing has been getting was probably around $35 in 1982, compared with the spot price low point of $17 per lb in November 1982.[21] This has implications for the future. The life of the mine is likely to depend much more on the market for uranium than on the size of reserves at Rössing. These are large — the present mining plan runs until the year 2020 — but low grade.

Rössing's view of the future

Like CDM, Rössing has been trying to establish an image in Namibia as a non-political company, interested only in the development of the country. As the Chairman put it, 'we will be prepared to deal with the new (i.e. independent) government. They only have to see our track record to recognise the company's importance to an independent Namibia'. Through the Rössing Foundation, an education centre has been established in Windhoek, with satellites at Okahandja, Okakarara and Ondangwa. A Rössing Foundation report proposed small rural development programmes in Ovamboland, Namaland and Damaraland bantustans, and in 1983 the Foundation announced that it would take up the Ovamboland proposal. The report expressed the belief that 'in the absence of an active department of agriculture (in Ovamboland) the Rössing Foundation Rural Training Centre, with its satellite centres and mobile teaching units, will become the major force in agricultural training and community development'. This seems unlikely, as the initial annual budget of R26 000 (after R75 000 capital expenditure) allowed only for one agriculturalist, who was required to double as administrator, and two part-time staff, a neddlework teacher and health educator/nutritionist. The 1982 report considered rural training in other bantustans, but rejected Kavango because of the excellent work of the ethnic administration, Rehoboth and Caprivi because of past difficulties or potential damage to the Foundation's image, and Bushmanland and Kaokoland 'because of logistical problems and the knowledge of the involvement of other agencies e.g., different church groups and the Defence Force'.[22]

Particular attention, as with CDM, is given to high level education, with scholarships to overseas universities provided both for potential mine management staff (what was called the Rössing

---

21. Dividing 1982 Rössing turnover (£218m.) translated into dollars at the average £:$ rate during 1982 by amount of oxide produced (4 910 short tons) would give $39 per lb. The 1981 figure would be only $31, however, and it seems likely that in 1982 more oxide may have been sold than was produced — no figures are provided for sales. Hence the estimate in the text. The NUEXCO spot price had risen to $22 by April 1983.
22. Information in this paragraph from *Guardian* 18 Nov. 1982, Rössing Foundation 1982, and *RTZ Fact Sheet* No.2, 1983.

cadetship scheme) and through the Foundation. Yet absolute numbers are small, and given the appalling level of school education it has proved difficult to find adequately qualified black students. 19 of the 33 students in 1983 were white.

Labour

Rössing employs about 3 200 workers. With CDM, it pays the highest wage in Namibia (a minimum of R296 in 1983), fixed according to a non-racial scale. Unlike CDM, it hires workers on permanent contract and it has a policy of providing family housing, although in 1983 one quarter of the unskilled workers were still living in single-sex hostels. The houses provided are of a reasonable size, with electricity and a water supply. This was not always the case: after international protests, the Chairman at the 1977 Annual General Meeting of RTZ admitted appalling housing conditions for those building the mine, and extra funds were allocated for upgrading. There are still clear divisions: unskilled workers (up to Grade 5) live in the new town of Arandis, built near the mine; a few semi-skilled workers (Grades 6 to 8) live in Arandis, but most live in Tamariskia, formerly the 'coloured' suburb of the existing town of Swakopmund; higher grades have another suburb, Vineta. In practice this means that almost all 'Africans' live in Arandis, almost all 'coloureds' in Tamariskia, and almost all 'whites' in Vineta. Rössing has also adopted a large training programme, with significant effect in promoting blacks into semi-skilled posts, though much less at higher grades. (Section 4.1.2 and Tables A9.1-4).

A sophisticated system of worker/management communication has been established, with elected worker committees for each area of the plant (albeit two in each area, one for higher grades and one for lower grades). There have been problems, and workers in lower grades have tended to choose representatives from higher grades rather than from among their own number, but management expresses cautious satisfaction.[23] Rössing says it does not see the committees as an

23. Rössing's Personnel Manager commented in 1983:

> In the last election 48 constituencies were vacant and only 22 of them were contested, the other 26 seeing employees nominated at the popular request of the constituents. The poll for the 22 contested constituencies averaged 93.6%, i.e. almost everybody eligible to vote actually voted . . . Only 26% of representatives are in grades 1 to 6 . . . Employees in grades 1 to 6 are choosing to elect employees in grades 7 to 11 to represent their interests, because they believe that those people are better able to do it. Last year in the Personnel division the elected representative for the grades 1 to 6 employees, who are almost all black, was a white woman.

Workers may also believe such representatives carry more weight with management — cf fn 40 below.

alternative to trade unions, and says it would welcome the emergence of a trade union with which to negotiate issues such as wages which are outside the scope of the committee system. Since 1980 no union has come forward. The explanation offered by the Personnel department is that unions have not been able to interest workers sufficiently — the implication is that workers are generally too satisfied to bother.

The workers' view

At this point, there is an alternative, almost diametrically opposed, perspective. Church and other sources report that black workers at Rössing, how great a proportion is impossible to judge, are angry, disaffected and bitter.[24] They believe that discrimination is rife, pointing to the attitudes of individual supervisors; to the very small number of blacks in senior positions; to whites and 'coloureds' living in Swakopmund (as skilled grades) whilst Africans live in isolated Arandis, close to the plant and surrounded by desert; to the allocation of changing rooms and buses;[25] and they believe management prefers whites and then 'coloureds' for training and promotion before Africans.[26] The training programme is claimed to be a joke or a public relations exercise, and what particularly rankles is the company's refusal to promote a qualified man unless there is a vacancy which he can fill: workers see this as a way of holding down black Namibians. The system of workers committees is seen as dependent on management's whim — workers have no real power within them.[27]

---

24. This attitude is not, of course, confined to Rössing. See Gordon 1977 p. 89 for very similar views at another mine.
25. Separate changing rooms are used by senior and junior employees, and more comfortable buses for the longer journey from Swakopmund than the shorter from Arandis. Company policy is that these are non-racial. However, most whites are in the senior grades (and hence Swakopmund) and most Africans in the lower (and hence Arandis).
26. Overall in 1982 promotions were in fact 27% of the African and Coloured labourforce, and only 17% of whites. The belief of these workers, however, is that in those grades where there is a significant representation of all races, whites and Coloureds seems to be preferred. Table A9.2 shows Coloureds — who generally have a better education than Africans — advancing into the skilled grades much faster than Africans. Similar accusations that Coloured staff receive preferential treatment have been made about the Rössing Foundation Educational Centre in Windhoek (WO 2 July 1983).
27. The Industrial Relations Superintendent at Rössing has a similar view, and even draws the same conclusion about the desirability of a trade union:

    A committee has no formal membership and therefore receives no subscriptions and has no economic basis. The initiative for the establishment of a committee usually comes from management, and the structure and rules of conduct are determined jointly by management and employees. These

The answer, these workers say, would be a trade union, but that they see as impossible, pointing to recent history. After two strikes in 1976, a branch of the National Union of Namibian Workers was organised in 1978. Arthur Pickering, a personnel officer at the mine, became the secretary of the union.

The NUNW found great hostility. In December 1978 the entire workforce went on strike, demanding not just higher wages but an end to racial discrimination, no police harrassment, and better housing, dust protection and recreational facilities. The police arrested the leaders and raided the workers' living quarters. Arthur Pickering was detained by the police three times, and eventually had to leave Namibia. By 1980 the NUNW nationally had been driven underground. Later in the year, some 200 of the Rössing workforce went on strike when a miner disappeared, apparently seized by the police. The company threatened to sack the strikers unless they returned to work immediately, said it could not be held responsible for police actions, but promised to investigate. Since then there have been no major strikes, and open union organisation has been regarded as hopeless.

Rössing and South Africa

Rössing management would no doubt regard all this as a gross misrepresentation of their position, but mine workers would say the same about management's view. Management believe they are winning round their workers to see the company in a positive light, but there is a fundamental problem. Despite all its efforts, the company is still identified in the eyes of its workers with the South African administration. This cannot be avoided, because Rössing cannot in practice isolate itself from political trends in Namibia. Two examples are telling. The company has admitted that it maintains a paramilitary force in close co-operation with the South African forces, in accordance with South African requirements for defending industrial establishments from 'terrorists'. Management has no choice: as the chairman of RTZ explained, the company is bound to try to protect itself in circumstances of civil strife. But it clearly signals to the

---

> differences enable a trade union to exert more pressure on management than a committee would be capable of doing, since its independent membership, backed up by its potential to exercise the strike weapon, enable it to deal with management from a position of equality. With its power base, the union can negotiate with management from a position of strength, whereas without a power base, employee representatives on a committee are only in a position to consult or offer advice to management.

Address to a conference on industrial relations 1982.

workforce which side the company is on. Furthermore, management cannot in practice escape by its own policies from injustice and discrimination in Namibia: African workers promoted to skilled grades which entitle them to move into Swakopmund have refused to do so, in part because schools for black children in Swakopmund are segregated and poor.

Workers at Rössing, as elsewhere, support SWAPO's campaign for independence. Whilst management is identified with government, there is bound to be conflict. Conversely, once the government is changed at independence, workers expect changes at the mine as well.

Health

Rössing says it follows stringent international rules in safeguarding the health of its workers. Management points to the fact that Rössing has a much lower amount of uranium in the ore than other uranium mines, and workers are in the open air rather than underground: the risk of exposure to radiation, the company says, does not arise before the final treatment plant, where workers are well protected.[28] Workers are regularly screened, and the chief medical officer plans a long study of the health of workers after they have left Rössing.

Other people are not so certain. In December 1979 a group of Rössing workers said: 'Working in open air, under hot sun, in the uranium dust produced by grinding machines, we are also exposed to the ever present cyclonic winds which is blowing in this desert. Consequently our bodies are covered with dust and one can hardly recognise us. We are inhaling this uranium dust into our lungs; many of us have already suffered the effect.'[29]

A 1980 United Nations hearing questioned Dr. Joseph Wagoner, who for 20 years worked on the health effects of uranium mining for the United States Public Health Service.[30] He gave detailed international experience, questioned the adequacy of international standards and concluded 'all phases of the uranium mining industry, from mining through to milling to waste disposal, have been, and continue to be associated with a major risk of adverse health defects and a tremendous risk of adverse ecological effects'.

In the case of Rössing, Dr Wagoner singled out the dangers of

---

28. Nevertheless, uranium oxide has apparently been stolen from Rössing. WA 8 Feb. 1983.
29. Quoted in SWAPO 1982. This is regarded as propaganda by the company, but only regular inspections by independent authorities could establish the true health risk.
30. UN General Assembly, UN Council for Namibia, *Report of the Panel for Hearings on Namibian Uranium* Part II (A/AC. 131/L 163 Part 2) 30 Sept. 1980).

waste being left uncovered, both because of wind blown dust and because of contamination of water. In 1982 the local *Windhoek Observer* newspaper gave prominence to both.[31] It printed a satellite picture of what it claimed was the dust cloud, reporting that scientific studies of the radon dust danger has led to responsible official bodies closing open cast mines in Canada and Australia. With reference to water, it reported uranium traces in water in the Swakop riverbed below Rössing, and questioned whether contaminated water was not also seeping underground, to be pumped up later by other users.

The *Windhoek Observer* reports were disputed. Dr D. van As, who visited Rössing to advise on environmental issues, commented on the thoroughness of Rössing's environmental protection by comparison to North America, and in contrast to Dr Wagoner wrote: 'Nowhere, except in underground mines or enclosed ore storage pads, has (radon) been found to build up to unacceptable levels'. The company argued that natural uranium deposits could be expected to produce traces of uranium in the water, and the Secretary for Water Affairs 'could state emphatically that everything was under control in regard to contamination'.[32] However, Rössing did find it necessary to drill 30 boreholes to recover water that had seeped from the tailings dam. Likewise, nobody disputed that dust from Rössing reaches some kilometres from the mine or that the town of Arandis is close by. In view of workers' concerns and the worldwide history of mining-related diseases being identified despite earlier company assurances, a newly-independent government might consider an urgent review of health at Rössing conducted by people whose authority is recognised both internationally and by the Rössing workforce themselves.

Opposition to Rössing

Rössing has become the focal point of opposition to foreign investment in Namibia, for two main reasons. First, the mine was opened with the full knowledge of the ruling of the International Court of Justice that South Africa's occupation was illegal; and second, Western power authorities, all of which work closely with their governments, are the principal customers. There is also the fear that Rössing uranium may be being used (along with South Africa's own supplies) in the development of nuclear weapons by South Africa.

SWAPO's opposition to the mine and its contracts has been loud. 'Mining of uranium in Namibia today is of no benefit to our people. On the contrary, the only benefactors of this illegal undertaking are

31. WO 6 Feb. 1982 and 16 Oct. 1982.
32. WO 6 Nov. 1982 and 30 April 1983.

the companies involved, the Western governments and the South African regime'. SWAPO 'regards the export of Namibian uranium as theft, and, as is provided for in Decree No. 1 of the United Nations Council for Namibia, SWAPO will claim compensation for it as the Government of an independent Namibia with the full authority of international law behind it'.[33] The UN Security Council declared that titles granted by South Africa to companies 'are not subject to the protection or espousal by their states against claims of a future lawful government of Namibia.' The General Assembly condemned 'the exploitation of uranium (in Namibia) . . . and demands that such exploitation, direct or indirect, cease forthwith' (Resolution 3394, Dec 1975). Churches too have opposed the contracts, and a resolution of the British Council of Churches is printed in Appendix 3.

Other uranium mines

Despite this opposition, Namibia was seen in the mid-1970s as potentially a rich source of profit from uranium. There was talk of Rössing's production being doubled to 10 000 tonnes per year. The South African-based Gencor group spent a considerable amount proving a possible mine at Langer Heinrich (possibly 3 000 tonnes p.a.). Anglo American, Minatome (France) and Omitaramines (France) identified another deposit at Tubas, and Gold Fields of South Africa thoroughly investigated one at Trekkopje. Other companies scoured more remote parts of the country. However, the collapse of uranium prices and the political uncertainty over Namibia's future have combined to delay decisions over whether to develop mines at any of these places.

Tsumeb Corporation Ltd

The mines

Tsumeb Corporation (TCL)'s operations centre round the base metal smelter and refinery built to process the extraordinary range of minerals in the Tsumeb mine.[34] To feed this processing plant TCL operate Tsumeb itself, three neighbouring mines (Kombat, Asis Ost and Asis West) and two mines near Windhoek (Matchless and Otjihase).[35] (Table 6). These mines were operated when mineral prices

33. For a recent expansion of SWAPO's position, see the speech by the Secretary for Economic Affairs in SWAPO 1982.
34. In addition to Murray, information on Tsumeb from TCL and Newmont Annual Reports, Tsumeb 1980.
35. Tsumeb also processes ore from mines owned by other companies.

Table 6

Mines Supplying Tsumeb Smelter — The Effect of Otjihase[1]

Mine	ORE DELIVERED 000 tons		METAL CONTENT[2] Copper 000 tons		Lead 000 tons		Zinc 000 tons		Silver 000kg		IDENTIFIED ORE RESERVES 1982 Positive Reserves 000 tons	Total Reserves 000 tons	*Years*[3]
	1981	**1982**	**1981**	**1982**	**1981**	**1982**	**1981**	**1982**	**1981**	**1982**			
Tsumeb Mine	494	385	17.8	12.5	36.2	25.7	10.3	7.1	49	35	3 500	6 251	*16*
Matchless	111	123	2.6	2.8	—	—	—	—	1	1	316	2 199	*18*
Kombat/Asis W/Asis Ost	311	292	9.3	12.4	6.0	5.6	—	—	8	11	2 169	3 843	*13*
Otjihase	207	769	3.2	15.0	—	—	—	—	1	4	4 093	9 975	*13*
Owned by others[4]													
Oamites	530	n.a.	6.0	n.a.	—	n.a.	—	n.a.	6	n.a.	n.a.	2 020	*4*
Rosh Pinah	1 920	n.a.	—	n.a.	6.8	n.a.	24	n.a.	n.a.	n.a.	n.a.	n.a.	*n.a.*
Klein Aub	n.a.	n.a.	5.9	n.a.	—	n.a.	—	n.a.	13	n.a.	n.a.	n.a.	*8*

Notes

1. The main change in 1982 was Otjihase reaching full production of 65 000 tonnes a month in April, increased to 80 000 from October, whilst from July mining was stopped in the lower section of Tsumeb mine in order to reduce operating costs..
2. Metal content calculated from per cent in ore mined and milled, except silver which is per cent in concentrate produced after milling. Oamites, Rosh Pinah and Klein Aub by varied methods.
3. Reserves divided by 1982 production figures.
4. Deblin also supplies the smelter (WA 25 March 1983), and there may be some from foreign countries.

n.a. = figures not available.

Sources:

TCL Annual Report 1981.

Klein Aub MAR. 1981 gives figures for 9 months to Dec. 1980. Life of eight years from ENOK 1981.

Rosh Pinah MAR.1981: annual production of 15 000 tonnes lead concentrate (45% lead), and 50 000 zinc concentrate (48% zinc).

Oamites MAR. 1982: actual copper production reported at 4 284 tonnes, but metal content assumed to be 1.13% of ore delivered, which is the grade of remaining reserves.

were high, and suspended when they fell — only Tsumeb itself was worked uninterrupted throughout the 1970s. They are underground mines, and produce not merely copper and lead, but also zinc, silver, cadmium, pyrites, arsenic and germanium. Preferential rates for smelting have also been offered to other mining companies operating in Namibia.

Ownership, investment and profit

TCL's principal owners were until 1982 the US multinationals AMAX and Newmont (with reportedly 29½% of shares each); BP Minerals International, formerly Selection Trust, (14%); and the South African GENCOR (9½%). Management was until recently handled by Newmont, and sales by an AMAX subsidiary. Sales have now passed to O'okiep, and the management team, all of whom have worked in Namibia for many years, are apparently allowed more freedom of action than before. Nevertheless, the TCL Annual Report is still issued from New York.

However, during 1982 the Anglo American complex of companies (which includes De Beers) took a major interest in Tsumeb, through Gold Fields of South Africa (GFSA). In October, after a rights offer, GFSA acquired 27½%, with only Newmont and BP Minerals International remaining unchanged: the other percentage shareholdings were effectively halved. Thereafter, AMAX sold its entire shareholding at what seems to have been a low price. By June, GFSA was reported to hold 42% of equity, and Newmont to have reduced its shareholding.[36] The significance of the move is enhanced by the 1982 attempt by GFSA's parent company in the Anglo stable, Consolidated Goldfields, to take over Newmont. The bid was beaten off, and Consgold agreed to limit itself to 26% of Newmont shares, but it is possible it will be renewed in future.

Chapter 2 showed the enormous profitability of Tsumeb during this century, and Appendix Table A6 lays out more recent performance. In line with experience worldwide, in the late 1970s profits fluctuated widely; profit after tax as a proportion of shareholders' equity was less than 1% in 1976, 63% in 1979 and negative in 1981. However, nearly all profits (93% between 1974 and 1981) are paid out as dividends to the shareholders and therefore leave Namibia, rather than, for example, being kept back for re-investing. Furthermore, tax, which averaged 39% of profits between 1970 and

---

36. *Daily Telegraph* 25 Apr. 1983, reports AMAX receiving $6m. for sale of TCL and O'okiep shares. Other information from RDM 17 June 1983 and TCL Annual Report 1982. In 1983 Newmont also reduced ownership of O'okiep to 49%, with GFSA coming in. (1982 Newmont Annual Report).

1974, has been minimised since 1980. This is partly because of the current crisis, but also reflects the purchase of the Otjihase mine.

Otjihase was a new mine, and an extension to the smelter at Tsumeb was built to smelt its ore. However, the then owners closed Otjihase in 1978 in view of the low metal prices. TCL bought 70% of the shares and restarted production in December 1980, so restoring the supply to the Tsumeb plant.

Labour in TCL's mines

TCL employed roughly 6 400 people at the start of 1982, 5 000 of them black, in its various mines (Table A10). Workers are largely on short-term contracts.[37] Tsumeb was one of the earliest mines at which strikes broke out in 1971 against the contract system, and there was an NUNW branch in the late 1970s. In an internal memorandum of 1975, the Anglo American Corporation criticised Tsumeb's low wages, which were half those on Anglo gold and coal mines in South Africa. Anglo also noted the lack of training for blacks, 'and the housing standards are primitive'.[38] Wages remain much lower than at Rössing or CDM — the minimum in early 1983 was about R125 per month, approximately half the CDM level.

Most black miners still have to live in hostels. At Matchless, workers live 12 to a room facing out onto a sandy courtyard, there is a communal eating hall with a television set, a place to brew and sell local beer, a football pitch, and a chapel. There have been some concessions to the workers' wish to be with their families. Transport between the mines and Ovamboland has been made more available, and a small number of rooms have been built at the compounds where families can stay when they visit. But the rigid labour structure remains intact.

TCL has a history of training Namibians but in the past they were almost entirely whites. The main area of change since 1975 has been the training of a group of black workers to semi-skilled and apprentice levels.[39] In 1982 there were 26 black apprentices and 49 white. Money was saved in some areas by replacing skilled by semi-skilled workers. Family housing was built for the new group — for 267 of the 5 000 black workers by 1982, although the housing programme for semi-

---

37. However, miners do return for additional contracts: only 15½% of the semi-skilled and unskilled workers changed during 1981, according to TCL Annual Report (4% in 1982).
38. Excerpts from these memoranda, produced when Anglo were contemplating buying AMAX's shares in Tsumeb in the mid-1970s, were revealed in Christie, n.d.
39. The Annual Report said 166 blacks were 'in jobs traditionally held by whites' in 1982, compared with 42 in 1980.

skilled blacks was suspended during the year because of the recession. TCL has also given money for schools in the Tsumeb area, but it has sought much less of a public image than Rössing or CDM.

The South West African Mine Workers Union has recently been recognised by management as organising the skilled labour level.[40] Otherwise there is no overt union organisation on TCL's mines. Nevertheless, the whole black workforce at Otjihase went on strike in April 1983 over the introduction of a new work regulation and the attitude of the Acting Mine Superintendant. Faced with a strike, management revoked the regulation and offered an investigation, but refused to suspend the superintendant. 112 workers who refused to return to work on this basis were considered to have resigned.[41]

Importance and future of TCL

TCL is not as important to its foreign owners as is Rössing or CDM. In its last year of profit (1980), it contributed only 3% to Newmont's profits in 1980 (10% in 1970), and less than 1% to AMAX's. To GFSA, however, TCL provides the centrepiece of a previously rather diverse set of holdings in Namibia. Through its subsidiary Gold Fields Namibia it owns the old SWACO company (p.53 below), which, in addition to a small equity holding in TCL, has a joint prospecting company with TCL, extensive prospecting rights and the right to mine a number of known deposits. GFSA also has the majority interest in Zincor, whose South African plant is the principal customer of the Rosh Pinah mine, and in the Trekkopje uranium prospect.

The future of TCL has been questioned in the past. Table 6 showed the Tsumeb mine as the heart of the operation in 1981, providing almost half the copper and 80% of the lead. Yet it is possible that Tsumeb mine is nearing its end, with both ore grades (Table 7) and recorded reserves falling.[42] The 1975 internal memorandum of Anglo American, drawn up when Anglo were contemplating buying AMAX's share in Tsumeb, said 'it is now almost certain that the mine has a finite life of about 13 years' (i.e. to 1988): there is no obvious reason why Anglo should wish to deceive

---

40. At Tsumeb, the union in 1982 had apparently recruited about 300 (out of 5 000) black members, who saw management as being prepared to listen to white union officials. Attempts by the SWA Mine Workers Union to organise at Rössing failed.
41. WA 22 April 1983. The 112 were offered re-employment provided they returned within 30 days.
42. The mine can of course decide both what grade of ore to mine in a particular period, and how much exploration to do in order to identify reserves. The figures therefore reflect mine decisions as well as the minerals left in the ground. Nevertheless, the trend seems clear.

Table 7

Falling Ore Grades at Tsumeb Mine

Percentage of Metal in a Tonne of Ore

	1973	**1974**	**1975**	**1976**	**1977**	**1978**	**1979**	**1980**	**1981**	**1982**
Copper	4.10	4.36	4.27	4.25	4.71	4.95	4.38	3.61	3.60	3.24
Lead	11.51	10.06	9.73	9.04	7.79	7.01	6.27	7.13	7.32	6.67
Zinc	2.65	2.30	2.47	2.39	2.04	1.68	1.66	2.26	2.08	1.84

Source: TCL Annual Reports.

itself. However, one cannot be sure that Tsumeb is approaching the end. The mine has operated for most of this century; its tentative reserves as reported in 1982 would last until 1998, if they could all be mined; more may yet be discovered; and its present ore grades are still better than mines elsewhere.[43] Just as important, TCL's purchase of Otjihase mine, with its large ore resources, gives a new life to the company. In 1982, Otjihase contributed more copper than Tsumeb mine itself. TCL remains a major operation on an African or world scale.

Other Mines

Ownership and mines

After the second world war, large international (including South African) companies came to control almost all Namibia's mines. Excluding the mines of the big three companies, there were seven other mines employing more than 100 workers in 1978, and all were owned by outside interests.[44] The largest were two copper/silver mines — *Klein Aub*, owned by the South African Gencor, and *Oamites*. Oamites and its small neighbour *Swartmodder* were developed by Falconbridge of Canada with the support of the South African state Industrial Development Corporation. Both Klein Aub and Oamites were mines with relatively short lives,[45] opened in 1966 and 1971 respectively. A similar mine, for tungsten, was opened at *Krantzberg* by US companies, Bethlehem Steel and Nord Resources Corporation in 1973. These were fairly small investments by international standards. Their profitability and whether they stayed open depended, as with the Tsumeb mines, on the price of minerals. Generally, the more succesful were opened before the price of base metals slumped in the mid-1970s.[46]

43.

Percentages of	*Copper*	*Lead*	*Zinc*	*Nickel*
Tsumeb production 1982	3.2	6.7	1.8	—
Tsumeb reserves 1982	3.6	4.2	1.1	—
BCL (Botswana) production 1979	0.8	—	—	1.0
Second largest US mine	0.7	—	—	—

44. Data on other mines from press reports, and sources in notes above.
45. Klein Aub estimated at eight years in 1981, and Oamites at four years (ENOK 1981, MAR 1981).
46. Oamites investment was Canadian $10m., shared with the South African state Industrial Development Corporation, and profits averaged $1m. a year until losses in 1981 and 1982. Klein Aub recouped its investment of R4½m. in two years. Otjihase, opened by JCI of the Anglo stable in 1975 at a cost of R42m. lost R6m. in the first year and was suspended in 1978.

Two mines with a longer history were the vanadium/lead/zinc mine at *Berg Aukas* and the *Brandberg West* tin/tungsten mine. Both were established before 1945 by the South West Africa Company (SWACO) which had been set up in Europe specifically to exploit Namibian minerals. By the 1970s, however, these two mines were also owned by transnationals. SWACO had passed into control of Goldfields of South Africa. The operations of the two mines reflected fluctuations in the world market, as well as falling grades of ore left in the mine.[47]

As we have seen in Chapter 2, the *Uis* tin mine and *Rosh Pinah* lead/zinc mine are both owned through subsidiaries by the South African state corporation ISCOR. Set up to feed South African plants, they have maintained production. They process more ore than any other base metal mine in Namibia, Rosh Pinah from its underground mine and Uis from open pit.[48] Uis's reserves are vast, but low grade; it produces tantalite and columbite as well as tin.[49] Rosh Pinah, whose production was increased 60% in 1976 to 1.9m. tonnes of ore per year, is less certain in the long term. In 1978, ore reserves were put at only 5m. tonnes, but all figures are highly suspect. ISCOR, which still receives 60% of its tinplate needs from Uis, clearly sees a future in Namibia: in 1982 it spent R¾m. on prospecting around Rosh Pinah, to be followed by R½m. around Uis in 1983. No figures are available on how profitable either mine is. Reported tax figures suggest that Rosh Pinah made a *taxable* profit of R5.3m. in 1982, and Uis perhaps R0.6m.; however the generous deductions allowable before tax is levied mean that accounting profits could be considerably higher.[50]

---

47. Brandberg West was closed from 1973 to 1976 as a result of low world prices, reopened in 1977 but finally shut down in 1980. Berg Aukas maintained production until 1978 when the mine was suspended as zinc prices fell. Processing of surface dumps was continued until 1982.
48. In MAR. 1981, Rosh Pinah was reported as processing 1.9m. tonnes of ore and UIS 0.82. Compare Table 6 for other mines. Rosh Pinah was said in 1983 to contribute R20m. p.a. to GNP (WA 18 May 1983).
49. Uis is another example of a large foreign corporation taking over and greatly expanding workings originally developed by smaller companies floated specifically to exploit Namibian minerals. Uis was worked on a small scale before the First World War, as were several other small scale tin workings. In 1958, however, they were all consolidated into the Namib Tin Mining Company, whose mining rights were then acquired by the ISCOR subsidiary IMCOR (WA 11 April 1980).
50. WO 2 July 1983 reports that Rosh Pinah paid R1.5m. tax in 1980, 0.7m. in 1981 and 2.1m. in 1982; at a 40% tax rate that implies profits of 3.8m, 1.7m, and 5.3m respectively. Uis was rumoured to have paid no tax until R0.24m. in 1982 (other information from RDM 25 June 1983, WA 18 May 1983). ISCOR's subsidiary also opened Windhoek offices in 1983.

Small mines

Of the small mines in Namibia, perhaps the most important are the Helicon and Rubicon of *SWA Lithium Mines*. Like the larger mines these too, having been established by small enterprises using hand-picking, were acquired by an international firm, Kloeckner & Co, and in 1968 partially mechanised and expanded. *Tantalite Valley Minerals*, owned by a South African firm, are thought to produce bismuth and beryl as well as colombite and tantalite, but with only seven black workers reported to the Chamber of Mines in 1980, production is unlikely to be significant!

Special mention should be made of *salt* production, which has apparently risen from 69 000 tons in 1967 to 235 000 in 1975 and 395 000 in 1981.[51] It is mined relatively simply from large deposits on the surface along the coast. About half comes from Salt and Chemicals (Pty) Ltd of Walvis Bay, which is owned 50/50 by two South African companies, Sentrachem and SWAFIL, themselves part of the giant Anglo and Barlow Rand groups respectively. The rest of the salt comes from a dozen firms with, most unusually, largely Namibian shareholders. About 90% of Namibia's salt goes to South Africa, where it has been reported to supply much of South Africa's industrial requirements. Salt has also been exported to Zambia, Zaire and West Africa. The reserves are virtually unlimited.

There are individual (white) Namibians active in the mining sector, as speculator-prospectors. The most famous is 'Giep' Booysen, said to be the discoverer of Otjihase and Oamites, former owner of the SWA Manganese Company and joint holder of a prospecting grant in the Aranos coalfield. But such men in practice also depend for their profits on large corporations. The prospectors speculate on likely areas, obtain the prospecting rights and then seek a corporation to sell them to. This was clearly shown when Charles Zandberg acquired the rights to an area next to Rössing.[52] He immediately flew to Johannesburg for negotiations. The only mine actually operated recently by Booysen was SWA Manganese at Otjosondu. That seems to have played a part in his 1982 bankruptcy, when Rössing refused to buy from it the manganese needed for uranium processing.[53]

---

51. WA 11 April 1980, 9 Oct. 1981, 8 June 1982; WO 14 Aug. 1982; *Namib Times* 10 April 1981, 17 Aug. 1982.
52. The area was taken back by the administration from an RTZ subsidiary because RTZ had failed to spend the required sum on prospecting it.
53. WA 14 Sept. 1982, 1 Oct. 1982; WO 4 Sept. 1982, 6 Nov. 1982. It is unclear how much mining was actually undertaken. Rössing rejected the manganese because it did not perform as well as its South African supplies. When SWA Manganese was auctioned by the liquidator in 1982, no companies came forward to bid.

Labour on the smaller mines

3 000 of the 18 000 miners in Namibia are employed outside the three main companies of Rössing, CDM and TCL. A third of these are at Klein Aub. At the smaller mines, like Deblin (p.25 above), conditions can be appalling. Salt workers complain of skin diseases. Even at the larger mines, conditions are generally worse than at Tsumeb. At Klein Aub, 934 of the 1 057 workers in 1980 were migrants from the north. A 1981 government report said they were given only simple training, and there was a 'squatter problem' because some of the migrants had brought their own families, and housing at the mine was inadequate.[54] A reporter for *The Lutheran* visited Oamites in 1982:

> The four-to-a-room housing for blacks is cramped . . . 'Coloureds', few in number in the Oamites work force, live on a separate site in houses afforded for them and their families. The white social director explained that the mine company would like to plant flowers in the turf between the rows of housing for blacks; 'But these people wouldn't know to take care of them', he says. I am also shown games, a recreation room and bar; 'they keep the men out of mischief', the social director says, . . . The health clinic has two entrances — one for whites and 'coloureds' where there is no line, and the other for blacks where half a dozen men are waiting.[55]

Oamites also provides an example of the distribution of income. In 1980 it recorded sales of R12.5m. (20m. Canadian dollars), with profits after tax of R0.9m. Its 52 white workers were paid R0.78m. (R15 000 each), 96 'Coloured and Rehoboth Baster' workers R0.66m. (R6 900 each), and the 265 Africans were paid R0.66m. (R2 500 each).[56]

3.4 Mineral Policy and the DTA

The voice of the transnational corporations

> The key to a healthy mining industry, and indeed to the wealth and welfare of the nation, will rest on the State's willingness to refrain from assuming control and management of the country's major business enterprises.
>
> Annual Report of Chamber of Mines, 1982.

Two voices were heard proposing mineral policy to the DTA

54. ENOK 1981 (Rehoboth).
55. *Lutheran World Information* May 1982.
56. Sales and net earnings from MAR 1981; wages from ENOK 1981 (Rehoboth).

administration established by the South Africans in 1978.[57] Not surprisingly, the most powerful is the voice of the large mining companies. It is heard particularly in the statements of the Chamber of Mines, and from the principal spokesmen of the individual companies.

The first thrust of the argument is the merit of free enterprise. Governments have the right to set a framework and to tax, but the state should otherwise keep clear. Private enterprise is more efficient and profitable, 'essentially because private enterprise management is more flexible and does not have to be responsible to the voters', in the striking words of the President of the Chamber of Mines.[58] State shareholdings and state intervention in management have combined with 'selfish local policies' to cause the steady decline of mining in certain independent African states.[59] The resident director of CDM even cited Botswana as following such socialist policies.[60]

Rather, the transnationals say that the state should hold the ring and no more; Gabon is held up as an example. The state should maintain laws which are not changed suddenly. There should be a sound framework of industrial relations. The three major companies say they are willing to accept trade unions within that framework, although as we have seen black miners in practise have found open organising impossible. The Chamber of Mines has also actively promoted amendments to the Wages and Industrial Council Ordinance along these lines.

The second theme in the companies' argument is the need to attract foreign investment. New mines will be needed, because Tsumeb and others are nearing the end of their lives. The capital and skill to build new mines can only, imply the companies, come from private foreign investment. Foreign companies will only come if they are allowed to take out their profits. Above all, they need to be allowed high profits in the first place, higher than other industries: 'Today's average (non-mining) investor expects a real return of 2½ to 4% per year above the rate of inflation . . . Mining investors have to look for a real rate of return around 15% above inflation, after tax'. Higher profits are needed, it is said, because mining companies take high risks on their huge investments: profits are needed to balance

---

57. The political history, which included internal elections boycotted by SWAPO, is almost as convoluted as the many-tiered administrative structure that was built. For detail, CIIR 1981.
58. WO 12 June 1982.
59. This and other quotations not otherwise indicated are from Mr Hoffe's speech as President of the Chamber of Mines to the Socio-Economic Conference, as reported in WO 14 Aug. 1982.
60. WO, 12 June 1982.

losses on the investments that fail.[61]

All this implies the state must do rather more than just hold the ring. It must construct a favourable tax structure, for example not taxing at all until profits have repaid the cost of the investment. It may even be necessary to subsidise marginal mines to keep them going when times are bad.[62]

The voice of local business

The transnational companies' view of their own merit has not gone completely unchallenged within the local establishment. In 1982 the maverick millionaire Eric Lang launched a blistering attack on the 'unholy alliance' between the administration and foreign big business. He accused CDM in particular of making huge profits, from which the state was not getting a fair share. Namibian resources were being over-exploited, and profits being removed. The government was hiding and distorting statistics.

Eric Lang is as much in favour of free enterprise as the large companies are, and part of his attack was on 'ill-conceived subsidies' as well as 'socialistic malpractices'. But his campaign fits with an unease among some local businessmen who see their livelihood threatened by unbridled foreign compeition. For example, in 1980 the *Windhoek Advertiser* complained that potential small mines were being held back by huge companies holding on to mineral rights and not making available the results of their prospecting. The unease spreads wider than just mining circles: in 1982 the mayor of Swakopmund called for Namibia to impose its own customs, because South Africa's customs duties hampered Namibian industries.

DTA policy

Between these two voices wandered the DTA administration. Whatever the economic arguments, foreign investment was essential politically, in order to gain respectability and legitimacy for both South Africa and the DTA itself. The DTA administration enthusiastically committed itself to free enterprise. 'In the spirit of this approach', said Dirk Mudge in his 1981 Budget Speech:

> This Government, in contrast to SWAPO, does not dictate to private undertakings with regard to employment practices, training and re-

61. Some of these issues are taken up in Chapter 4. See also Brown and Faber 1977 for counter-arguments.
62. In 1978 an electricity subsidy was introduced for a year following representations by the Association of Mining Companies of South West Africa (predecessor to the Chamber of Mines) on the crisis facing base metal mines. The Association also had a Marginal Mines Sub-Committee.

investment of profits . . . I have the fullest confidence that the investment and the prospecting activities currently undertaken by these firms will yield high returns for them in the long run, and this Government will not begrudge them that.[63]

In its enthusiasm for the free market, the administration even set about removing subsidies that gave advantage to Namibian producers over their competitors. The Chamber of Mines was not pleased.[64] In general, however, mineral policy followed the companies' line. In early 1981 the administration noted the need to subsidise marginal mines in order to create jobs and raise the growth rate.[65] After representations from the Chamber of Mines, mines were in 1982 given exemption from General Sales Tax on certain items, to bring them in line with South African mines. At the Turnhalle Conference itself (which gave rise to the DTA), Dirk Mudge set up an economic advisory committee. Its eleven members were all drawn from the top of South African and Namibian business, and included three prominent mining men. The Committee's report dealt extensively with mining, and as might be expected endorsed the companies' position. The state's main role was to attract expert mining companies.[66]

The most complete statement of free market policy came from the state investment corporation, FNDC/ENOK:[67]

> Low taxes are levied on profits. Under normal circumstances a new mine may recover all capital expenditure out of profits before starting to pay tax. And when it does, the tax is at the ordinary rate for companies, except in the case of diamond mines. There is no compulsory participation by the state or by local interests in mining ventures, no obligation to plough back profits or recruit local managerial personnel, no pressure to process minerals in the country, and no restriction on output volumes.

As this statement makes clear, the government refuses to intervene to create a Namibian mining sector, to stimulate new industries around mines, or to decide how fast Namibia's minerals

---

63. Quoted in SWA, *On The Economic Front* no.5, Nov. 1981, which expands on the theme.
64. The 1981 Annual Report complained of a reduction in capital redemption allowances and withdrawal of export incentives, and chided 'we look to Government to assist when it can in ameliorating some of our difficulties'.
65. SWA, *On the Economic Front*, no 3 (Jan. 1981) p. 5.
66. The mining men were the executive chairman of Rio Tinto South Africa, the director of Rössing, and a past president of the South African Chamber of Mines. The committee did also say that the state should check whether minerals were sold at fair prices, and see that mines did not just dig the richest ore, but there was no obvious response to this advice.
67. ENOK 1981.

should be dug up. By taxing all except diamonds at ordinary company rates and taking no significant other revenue, it ignores the possibility of some mines being especially profitable.[68] Instead it goes all out to attract foreign investors, without insisting on a negotiated code of investment conduct which will serve the long term interests of Namibians.

Change

In time the DTA did listen more to the interests of local business. The law governing prospecting licences was changed to prevent large companies holding on to deposits without developing a mine. In August 1982, in a move to try to re-establish his political support, Mudge announced what was seen as a radical change. His administration was considering as alternatives: state shareholdings in the major mining companies; state representatives on their boards; and stronger government departments to oversee the industry. No further detail on these alternatives emerged before Mudge's resignation five months later, so it is difficult to speculate on what, if anything, the DTA hoped to achieve.

The specific inadequacies of these policies will become clearer by comparison with the analysis in the next two chapters, but there are two fundamental problems. First, there is no way that the DTA, depending in the eyes of the people on the occupying military power of South Africa, could act genuinely in the name of the people. Mineral resources are recognised as belonging to the people of a country, and the DTA simply did not represent the people. Second, the policy of foreign investment at all costs ignores Namibia's interests. It is true that an independent government dealing with transnational corporations will have to consider the profit requirements and other minimum demands of those companies it wishes to stay. But partnership with the companies is likely to be acceptable only so long as it contributes to longer term goals.

3.5 The Present Crisis

The Overall Crisis

The picture that emerges in the 1970s is of a mining sector vital to the

68. Income Tax stands at 50% on diamonds, and 40% on other companies. In fact, the relevant Ordinance does allow for government to impose taxes or rent in making a mining grant, as Ushewokunze points out. But there is no evidence that this provision has ever been used to a significant extent. Uranium taxation is unclear, unless an ambigious provision in the 1981 Income Tax Act means that uranium mines are taxed like gold mines.

income and stability of the regime in Namibia, but owned by transnational corporations and very exposed to changes in the world economy. From about 1980, however, this careful construction was dramatically rocked. The mining sector is now in crisis. The reasons are the sharp slump in the world economy, and the underlying doubts of transnationals about how long the present regime can survive.

Figure 3
Boom and Slump

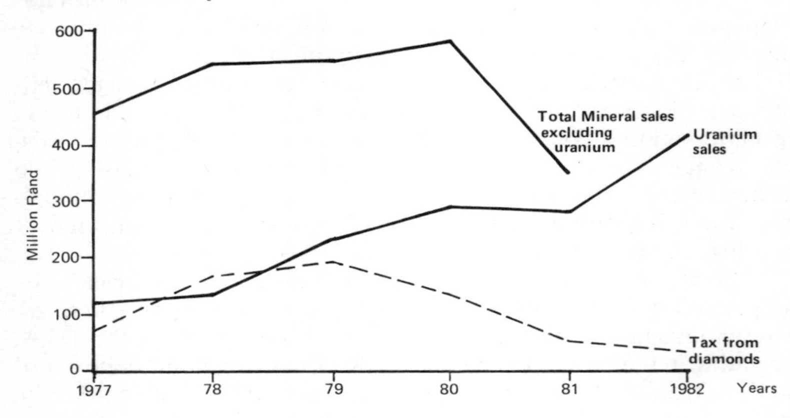

Note: 'Tax from diamonds' is the estimate for all revenue from diamonds. Revenue from other minerals was comparatively insignificant.
Sources: Tables A3 and A8

The slump is clear in official statistics (Fig.3 and Tables A2 and A3). The slide began in 1979, but the major fall was two years later. The value of minerals produced in the year to 1 October 1981 was 38% less than a year earlier in money terms. In real terms, it fell 9.8% in 1981 and 7.5% in 1982, according to the Administrator-General's budget speech. With increasing interest rates on debt, profits fell even more, meaning less for the mining companies and much less for the government to tax.

1982 saw an overall 14.7% improvement in sales, largely because depression in South Africa pulled down the value of the rand — the physical volume of base minerals rose by 7½% over the year, but they earned 30% more rands. Nevertheless, profitability continued to fall, according to the Chamber of Mines. Government received negligible revenue from the non-diamond sector, and diamond revenue, its main source, was reduced to a trickle — 7½% of expected income in

1982/83 compared with 42% two years earlier. Despite signs of improvement, it was still 7½% in 1983/34.

The crisis for each mineral

Diamonds

In 1981 the world's speculators decided that diamonds were no longer a good investment.[69] Despite continued demand for diamond jewellery, the loss of this investment market was extremely serious. De Beers as controllers of the world market acted to hold up prices by cutting sales. In 1981 they offered 42% fewer diamonds than in 1980. As always, the CDM mine was used as one of the principal regulators, with sales down 47%.[70] Between 1980 and 1982 the number of carats produced was cut by 34%, and the amount of ore moved by 40% Even so, perhaps a third of CDM's diamonds were stockpiled rather than sold. One of the four main processing plants was suspended, along with a number of other plants. Costs were temporarily reduced by mining richer areas than before (Appendix 4), and by stripping much less of the overlying sand in preparation for future mining — 18m. tonnes in 1982 compared with 57m. two years earlier. Despite these savings, CDM reported net profits down from R117m. in 1980 to R35m. in 1981 and R28m. in 1982. 1983 saw a modest revival in the world diamond market. The effect on CDM profits, however, will be muted by a reduced share in world sales (as production expands elsewhere).

Base metals

World slump is quickly reflected in falling metal prices, and this had immediate effects in Namibia.[71] In 1978 Berg Aukas, Otjihase and the small Onganja copper mine were closed. In 1980 Krantzberg and Brandberg West joined them. In 1981 Swartmodder was suspended,

---

69. To oversimplify, the reasons were these. In the late 1970s, the interest rates paid to investors by banks were less than the rate of inflation. Short-term speculators moved to buy diamonds, in the hope that the price of diamonds would go up faster than inflation. By 1981, however, bank interest rates were back above the rate of inflation, and diamonds looked less secure. The speculators moved out.
70. Table A2. At the same time the Lesotho mine was closed, as was Koffiefontein in South Africa, and 12% of workers were dismissed at Premier mine in South Africa — altogether 2 500 jobs went. Botswana, however, escaped production cuts, and most of its investment also went unscathed. Observers concluded that De Beers saw Botswana as the future jewel in its crown (*Financial Times* 4 June 1982). On the other hand, Botswana was clearly being required to stockpile rather than sell its share of production. Its diamond revenue in 1981 was 40% down on 1980. (RDM 3 June 1982, FT 15 May 1982, 18 May 1981).
71. In 1981 copper prices were at their lowest level for 30 years, in real terms. Since then, the price has recovered somewhat.

less than a year after starting production. In 1982 Falconbridge sold Swartmodder and Oamites, after a loss of 0.9m Canadian dollars in the year. The Kiln Products operation to process Berg Aukas's waste dumps was closed early, after being said to be losing R1m. a year. Some of these mines could reopen if prices improve, and indeed TCL has recommissioned Otjihase against the trend. But at others, such as Brandberg West, the equipment has been dismantled.

Tsumeb Corporation itself temporarily suspended three of its mines in the late 1970s in response to low copper prices. By 1981 they were all back in operation, but sales were 32% down on 1980. TCL reported losses of R3.4m. in 1981 and R8.8m. in 1982, despite a recovery in sales. Management responded by cutting expenditure, mining higher grade ore, raising loans and making a rights issue.

Even salt producers were in dire trouble, although for a different reason. In 1981 South Africa declared Walvis Bay an industrial decentralisation area within South Africa, as part of its policy of excising it from Namibia. Industries operating there were allowed special subsidies. As a result the Walvis Bay salt mines were able to undercut the prices of other Namibian salt producers.[72] However, after urgent representations from the Chamber of Mines, the Walvis Bay subsidies were rescinded in October 1982. That in turn led the principal producer at Walvis Bay to seek renewed assistance, on the ground that their South African customers 'are able to import salt from Australia, despite the long haul, at a lower cost than we are able to deliver from Walvis Bay'.[73]

Rössing to the rescue

The major exception to the picture of gloom is Rössing Uranium. Its profits rose, and in 1982 pre-tax profits were considerably more than double those of CDM. With the collapse of diamond revenue, the start of tax payments by Rössing in 1983 comes as a godsend to the administration: R30m. was expected in tax from non-diamond mines in 1983/4, all but R2m. of it from Rössing.

Yet Rössing's profits are far from secure. The uranium market too is very depressed and likely to remain so. As we have seen, Rössing is still very profitable because its prices were fixed in long term contracts. But the contract prices are now well above the short-term 'spot' price — and contracts do not last forever. Early in 1983, the

---

72. The major incentive was a 10% transport subsidy, which is significant because transport makes up 75% of the cost of salt delivered in South Africa. Much higher subsidies are also possible under decentralisation rules. WA 9 Oct. 1981, 8 June 1982; WO 14 Aug. 1982. *Namib Times* 17 Aug. 1982. Chamber of Mines 1982. FM 22 July 1983

73. Chairman of SWAFIL, quoted in RDM 18 May 1983.

British Central Electricity Generating Board announced that it would not renew its contract. The problem has been neatly summarised by the *Financial Times:*

> Rössing is a low grade mine and thus has to be worked on a large scale to be economic . . . It is necessary for the mine to cling to its sales contracts in order to avoid production slipping back to levels at which the operation becomes uneconomic . . . In the present situation of heavy over-supply, Rössing has to face the possibility of receiving less-favourable renewal prices. Nor can it be easy to obtain new contracts in the face of heavy competition . . . Existing contracts are reckoned to keep the company going for the next five years. By then, it is hoped the uranium market will have started its recovery.[74]

Other observers doubt the recovery will come so soon. The uranium market is heavily over-supplied: in 1980, 44 000 tonnes was produced worldwide but only about 30 000 tonnes used. Mines were developed to meet an expected boom in nuclear power plants after the 1973 oil crisis. But the boom never came. It was overtaken by an economic slump, falling oil prices and worries about the safety of nuclear power. Some mines are being closed or cut back, but other new low cost mines are still being opened elsewhere in the world. Despite restrictions on mining by the new Australian government, the market is likely to remain over-supplied until 1990 and quite possibly until 2000.[75] At present the electricity generating utilities in the West have such large uranium stocks that some of them have been selling uranium themselves. Some new customers may be found, including Taiwan. But Rössing's golden eggs may be fewer than was thought.

The war

If the world slump is closing mines, it is the war that is preventing new ones opening. Such a statement needs qualifying. It would be difficult to make a new mine pay in the current recession, and if mineral prices boom transnationals would be tempted to invest again despite the war. Nevertheless, as the Chamber of Mines delicately put it, 'There is no doubt that the current political uncertainty contributes to instability within the industry . . . Prospecting continues at a relatively low level and the Chamber remains convinced that uncertainty regarding the political future of the country is the main contributory factor'.[76]

74. Quoted in RDM 3 Nov. 1982.
75. OECD 1982. See also Secretary-General, Uranium Institute, as reported in RDM 26 Mar. 1983. For an opposite view, RDM 20 Apr. 1983.
76. Chamber of Mines 3rd Annual Report 1981. The 1982 Report expressed the same sentiments. as, more generally, did the Private Sector Foundation (WA 4 Apr. 1983).

The extent of the war, the behaviour of the South African occupying army, and the suffering of the people have been documented elsewhere.[77] The army goes everywhere, and even searches mine hostels, at Tsumeb sometimes three times a month. Army personnel are also reported to use mine housing at the suspended Berg Aukas mine. However, military clashes are largely confined to the North. Mining has not been disrupted, with the exception of a sodalite mine in Ovamboland which was closed in 1982, according to its manager because of SWAPO activity. Prospecting in the North has been more seriously affected, and the Etosha oil exploration in particular is said to be impossible until the war ends.[78]

Holding back

The greater fear of transnationals is that any new investment might not be safe in an independent Namibia. Existing companies stress that they are 'apolitical' and 'good citizens', hoping for business relations with a new government[79]. Nevertheless companies must be aware of international law and of SWAPO's declared intention to seek compensation from companies who have broken it. Prospecting does continue, with R15m. spent in 1982. But Dirk Mudge was no doubt correct in accusing firms of using prospecting licences to hold on to reserves rather than to open new mines. As the 1981 *Mining Annual Review* put it, 'The development of Namibia's other identified uranium deposits has remained in abeyance pending a resolution of the country's political status . . . In particular, it is thought that Anglo American does not wish to invest in a new mining development prior to independence'. Likewise, the South African state corporation SOEKOR has had no success in attracting an international oil company to investigate the gas field off the mouth of the Orange River, and attributes this to political doubts over Namibia.[80] It remains to be seen whether the 1983 reimposition of direct South African rule over Namibia will make transnationals more confident.

The impact of the slump on Namibians

The most immediate effect of the slump was on the workers involved. Some lost their jobs: there were at least 2 000 people employed on the mines shut between 1978 and 1982.[81] Elsewhere, rather than being dismissed, workers were simply not replaced when they left. In July

---

77. CIIR/BCC 1981; SACBC 1982
78. WO 11 Sept. 1982, WA 7 July 1982.
79. For the use of 'apolitical', Chamber of Mines 1980; for 'good citizen', RDM 3 Nov. 1982.
80. *Financial Mail* Energy Survey, 13 May 1983.
81. Including Otjihase, but no other TCL mines.

1982 Rössing announced that it was not replacing any of the workers who left each month; by February 1983, there were 175 fewer workers. TCL too said that it was not filling any of its 250 vacancies, and by the end of 1982 had reduced employment by 175 workers. The migrant labour system offers CDM a peculiarly Namibian way of cutting the workforce: at the end of their short-term contract, miners are sent home and told to wait until they are called back to work. As a result, the total number of workers at CDM at any one time seems to have fallen from 5 800 at the start of 1982 to 5 200 in September 1982.[82]

Miners were not the only ones to suffer. With little tax coming in, the 'Ministers' Council' suspended all new development projects in 1982, and the subordinate 'ethnic administrations' were also cut back. In Ovamboland, for example, people found there were far fewer jobs on road-building or pipeline construction. Young men who had never been to work found there were no jobs, and were forced to stay in over-crowded 'homelands' or unemployed in town.

The end of the DTA

If the slump hurt ordinary Namibians, it was also one of the nails in the coffin of South Africa's effort to produce an alternative to SWAPO. In practice, the DTA administration never seemed credible in the eyes of the people. Unrepresentative, it was tainted, like the Vichy government in Nazi-occupied France, by collaboration with the occupying power. Falling revenues merely emphasised the point. The DTA could scarcely appear independent when three-quarters of its income depended directly on the South African Government.[83] As money became tighter, the financial extravagances and scandals surrounding the 'ethnic administrations' became more and more embarrassing. Spending had to be cut. The economic strategy of cutting taxes to boost free enterprise had to be reversed. For these and other reasons, popular cynicism rose and in January 1983, Dirk Mudge resigned. The DTA administration was at an end.

The finances of the South African administration remained a nightmare to its own civil servants, trained to encourage the free market, private enterprise and balanced state budgets. A confidential memorandum of the Department of Finance leaked to the press in June 1983 noted that government expenditure, because of the slump,

---

82. See footnote 12 above. The De Beers Annual Report showed a much smaller fall, from 6 765 to 6 541, in total employment. But this figure presumably includes migrants at home between contacts. In July 1982, CDM said the average tour of duty of migrant workers had been reduced from 8 to 7.6 months.

83. 23% direct grant, 30% customs revenue passed on by South Africa, 21% loans guaranteed by the South African government. (Estimates 1982/83). In 1983/4 the figures were 25%, 26% and 20% respectively.

would be 62% of GDP in 1983. Meanwhile state revenue was falling, and the administration was having to borrow — the memorandum expected Namibia's debt as a proportion of national income to surpass Zaire's in 1985 and to be equal to the entire national income in 1987. It concluded that 'expenditure is already moving beyond the financial means of the central government and it is expected that within a few years it will be completely out of control . . .' Particular criticism was reserved for the ethnic authorities:

> The present constitutional dispensation, whereby Representative Authorities are allowed to feed like parasites on the fruit of the land without any control or supervision and without delivering any corresponding returns, is fast leading to a collapse of the whole country. . . . Uncontrolled expenditures and shortfalls on the budgets of these authorities simply have to be supplemented (by the central Department of Finance) in order to allow these administrations to survive.[84]

Independence for Namibia still seems as far off as ever. Nevertheless, it is still important to examine the kind of mineral policies which an independent government might follow as part of a development strategy designed to meet the needs of all Namibians.

---

84 WA 22, 23 and 24 June 1983. The 'Coloured' administration, for example, exceeded its 1982/3 Budget of R15.5m. by R8.2m. Overall, the 1983/84 central budget had to increase direct grants to ethnic administrations from R211m. to R285m., plus an extra R25m. for health and welfare, and pay off R23m. of debt contracted by these administrations (WO 11 June 1983).

4 Mineral Policy Issues After Independence

Basic issues

Namibians will determine their own mineral policy after independence; to assist their decision-making, the United Nations has already undertaken at least three studies. This booklet therefore has no intention of making firm or detailed proposals. Rather, it seeks to outline the constraints imposed by the past, and the basic issues confronting an inheriting, popularly elected government. This chapter considers policy within the mineral sector; the next chapter examines the role of the mining sector in overall development policy.

The inheritance is crucial. It is an inheritance of lack of control by Namibians. The existing mines are operated by foreign corporations (or their subsidiaries). Black Namibian miners have been kept almost entirely unskilled or semi-skilled, at the bottom of a hierarchical structure. The foreign companies organise the sales of minerals. They also provide the technology required to run the mines. In the interests of profit, companies may run the mines 'efficiently'. But it is efficiency from the point of view of the company, not of Namibia. A corporation may decide to concentrate attention away from Namibia on Australia, say, or Canada, because 'conditions in Africa are too risky' — the EEC Commission has drawn attention to this trend.[1] To take another example, if the diamond market improves, it is at present a matter for De Beers alone whether to increase production in Namibia, or elsewhere.

To meet this inherited lack of Namibian control, independence is likely to bring a much more active role for the state. This is in line with

1. According to Mikesell 1979 14 European firms who were spending 57% of their exploration expenditure in Third World countries in 1961 spent only 11% of it there in 1976. Brown and Faber 1980 and Radetzki 1983 argue, however, that the position is more complex than this statistic suggests.

experience worldwide. There is widespread international acceptance of the right of peoples to the minerals under their soil. In 1962, a United Nations General Assembly Resolution declared:

> Violation of the rights of peoples and nations to sovereignty over their natural wealth and resources is contrary to the spirit and principles of the Charter of the United Nations . . .[2]

This is accepted worldwide to mean that the government not only decides the direction and purpose of mining in its country, but also plays an active role. Figure 4 reproduces a Canadian statement of aims of mineral policy as quoted with approval in a World Bank publication. Brown and Faber have argued:[3]

> there would be a wide measure of agreement in rejecting a role for the State which was purely regulatory, i.e. a matter of controlling actual or potential abuses in the context of a free market system. The desire of the developing countries to play a more active part, to be participants in the development of major projects, is now an established feature of the climate of business . . .

SWAPO's policy framework

Decisions on what policies to follow are clearly a matter for Namibians themselves. It is generally agreed, even by the South African army, that the liberation movement, SWAPO, has widespread popular support, and would win a free election in Namibia. Its views therefore offer the best guide to the policies of an independent government.

SWAPO has said that it wants to encourage mining. The long term aim is that mines, like other major industries, should be publicly owned and operated, rather than owned by foreign transnational companies. However, there is likely to be a considerable period of transition. Full independence from foreign enterprises is blocked in the short term — even if it were wanted — by the inheritance of the past. Machinery, markets and above all skilled people must be found abroad. If the present companies withdraw or are found unacceptable then alternative foreign enterprises may be called in. But dependence on foreign partners is inevitable. SWAPO itself has indicated its willingness to deal with Western mining companies after independence, on the basis of mutual respect for each other's

2. Resolution 1802 of 1962, extracts of which are printed in Brown and Faber 1977 p. 19.
3. Brown and Faber 1977, p. 2. Daniel 1983 deals with the international role of governments.

Figure 4
Canadian Mineral Policy Objectives

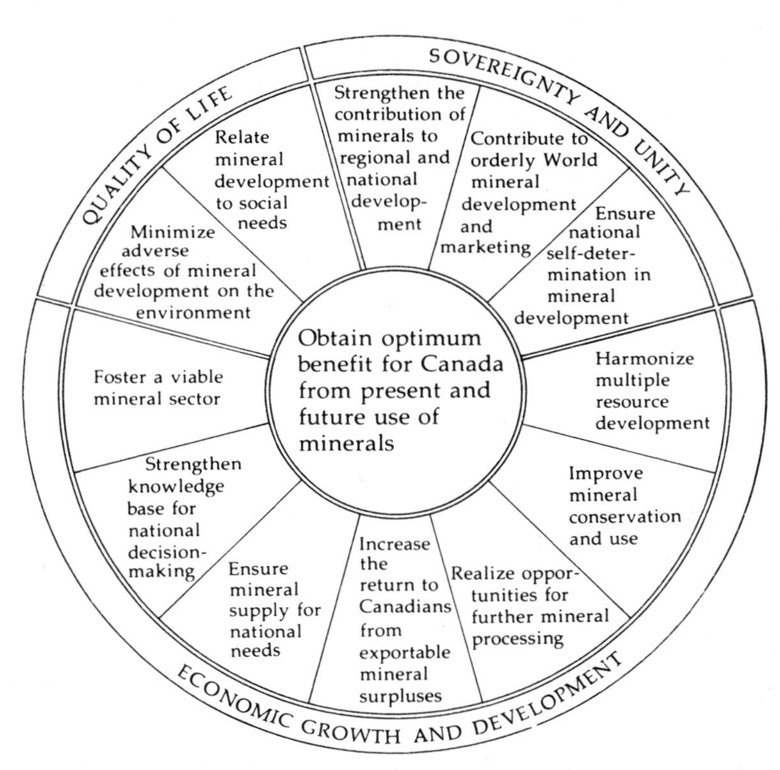

Source: Bosson and Varon, 1977

interests.[4] SWAPO has deliberately not prejudged negotiations by going into detail about possible arrangements.

The major issues for the future

The issues confronting a newly independent government therefore fall into two groups. On the one hand the more short-term issue is to take control: to set a framework for mining, and then within that framework to reach acceptable and stable agreements with foreign mining enterprises. Government goals are likely to include maximising

---

4. Sam Nujoma, President of SWAPO, speaking in London in late 1982. See also Chapter 5 below.

revenue and stabilising its flow; ensuring acceptable labour practices; expanding prospecting and collecting all data on mineral resources; obtaining ultimate control of the speed at which mines are worked; and generally ensuring fair dealing. The various different possible relationships with foreign parties are examined below. At one extreme the state might own a mine and employ foreign management; at the other extreme, foreign companies might have a majority of the shares.

On the other hand the long term, strategic issue is to build up Namibian participation: acquiring skills, including the ability to understand and judge technology; becoming accepted in, and knowledgeable about world markets; and bringing popular opinion to bear, for example through worker participation in some form, so that 'Namibianisation' is not a process whereby a small number of Namibians take over the benefits of a small number of outsiders. We turn first to this.

4.1 Laying the Foundations for the Future

1 Creation of a national mining service

The legacy of the past, Namibians' own long term goals and experience elsewhere in the world all point in the same direction — to the great importance of building up a strong government mining service. Whatever decisions are made about ownership, Namibia's interests can only be ensured by a knowledgeable and competent mining service independent of foreign companies and fully aligned with Namibia's wider, long term development aims. A state mining service would have at least five roles:[5]

(a) It is essential for *applying the government's policy*. Expertise is needed so that the long term aims of the new government can be translated into realistic programmes.

(b) It is essential to *fair relationships with foreign partners*. No relationship can survive the inevitable conflicts and suspicions if the government has no means of checking whether it is being duped. This will require the services of mining engineers, geologists and financial analysts, and personnel with a knowledge of markets. It also calls for a detailed reporting system from mines to the mining service.[6] For negotiating new agreements

5. For more detail on the organisation and manpower needs of a national mining service, see Zorn 1978 and Aulakh and Asombang 1981.
6. For detail on monitoring systems, see for example, Lipton 1978. An example of the kind of issue to be watched is the grade of ore mined — see Appendix 4.

lawyers will also be needed. Short-term assistance can often be obtained on specialist matters from international organisations.

(c) It is essential to *development planning* for the whole country. A government is severely handicapped if it does not have some idea of future mineral revenues, and of the likely life of the various mines. This requires information from the mines, and expertise to interpret and assess it. Predicting the future of mines also requires forecasts of market conditions, where co-operation with other countries (or independent analysts) can be useful.

(d) A strong state mining service is essential for the *best exploration of the country's mineral resources*. The present law allows foreign companies to hold prospecting licences for long periods without having to carry out much exploration.[7] This encourages companies to hold on to specific areas without developing them, either to deny the area to competitors or in case price rises or other changes make the deposit more profitable. Companies also pass on few data to the Geological Survey. Conversely, the present Geological Survey has no field prospecting capacity of its own. A new government will probably require a more active national exploration strategy. This strategy would be aimed in part at increasing knowledge of the basic geology. But it would also concentrate on areas which have the greatest potential, and on areas where there are specific needs.[8] The Geological Survey probably will not undertake all prospecting itself, with some areas offered to private companies. However, in that case, a new legal framework will be called for, where companies prospect specific areas under a work programme agreed with the government, give up the areas after a set period of time, and supply the Geological Survey with detailed information as prospecting proceeds.

(e) Perhaps most important, a strong government mining service is likely to be at the centre of *long term policy for gaining Namibian control of the mining industry*. It is the logical place to build up Namibian investment in mines.[9] A well-informed mining service can administer the state's shareholdings in joint-venture mining

---

7. In 1982, companies were supposed to spend R6 000 per year on exploration for each 100 km^2 they held.
8. Such as clay for brickmaking. Though both goals and methods may differ in Namibia, it may be valuable to compare the process of drawing up a strategy with Botswana's programme for geological exploration of the Kalahari (Jones 1979 pp 60-64). See also D. Green 1983 on lessons from Zambia.
9. The theoretical alternative of selling shares to individual Namibians would not meet the objective of spreading the benefits of mining to Namibians as a whole, if only because the Namibians who could afford to purchase such shares would be the few who are already wealthy.

companies, providing advice for government directors and planning for an increasingly Namibian mining sector. In addition, it can provide a focus for accumulating Namibian expertise in mining. The number of reliable and committed Namibian mining engineers and other professionals will be small at independence and for a considerable time thereafter, because of the length of training and experience needed.[10] Building a central core of such people within the state mining service as well as building up groups of Namibians in a few operating mines is more likely to contribute to the long term development of the mining sector in the interests of Namibia than if all are absorbed as isolated individuals into the present structure.

State mining corporations?

At the centre of the state mining service will inevitably be a ministry, with supporting technical departments. The new government may also consider establishing a state mining corporation, as a body outside the formal civil service restrictions.[11]

There are strong arguments for not establishing such a corporation simply to hold government's shareholdings in joint venture companies. On the one hand, it would involve duplicating tasks already being carried out by the mining ministry, and that would appear an unnecessary burden on the very few skilled Namibians available. On the other hand, the state's interests extend beyond simply mining issues into areas like health, protection of the environment, physical infrastructure, housing policy and overall state revenue requirements. A mining ministry, in consultation with the other relevant ministries, is perhaps better placed than a state corporation to deal with such a range of responsibilities. A central core of Namibian staff could also be built up within a ministry.

At the same time, a government ministry is not designed as the sort of body that can undertake the day-to-day management and operation of a mine. A separate mining corporation, with its own accounting procedures, is undoubtedly more suitable. If the mining

---

10. A few are already in training outside the country, under SWAPO's programmes. There will of course be professionals working in the existing companies prepared to regard themselves as Namibians. To the extent that they are prepared to work with the policy of the democratically elected government, they will presumably be welcome.
11. There is no existing state organisation in Namibia to build on: the (always) small interests of the old Bantu Mining Corporation were taken over by the state-run First National Development Corporation (FNDC/ENOK) and further reduced. The South African state IDC and ISCOR holdings are administered from outside Namibia.

sector is increasingly to become a genuinely Namibian operation through the route of state involvement, the state will increasingly take up such an active operating role in mines. Here, decentralisation may be useful, and one possible eventual model is for a number of corporations, for different mines or groups of minerals. Expert opinion is divided on whether it is also desirable to establish a central co-ordinating state mining corporation: proponents argue that there are issues of day-to-day operation which affect the entire mining sector or more than one group of minerals; opponents that it would spread scarce skills too thinly.

What are the routes to such a model? An important requirement is that Namibian professional staff acquire practical operating experience. One possibility is simply to use the restructured joint venture mining companies themselves, appointing Namibians to work with the expatriate management provided by the foreign partner: the example of Sierra Leone's arrangements with Selection Trust was commented upon favourably in a United Nations Institute for Namibia draft. Over time, Namibians would take an increasing role in actual operation, and the joint ventures would become increasingly like operating state mining corporations with foreign technical and financial assistance.

An alternative, incidentally the route followed by the British National Oil Corporation when it acquired Burmah Oil's operations off the coast of Britain, is to concentrate the effort, and more immediately to take over an existing mine or mines as a base for a new state corporation. The corporation would actually operate the mine, hiring staff as necessary, with a reduced role for any foreign management partner. Candidates mentioned include Rössing, in view of its importance and the illegality of its present operations, and Tsumeb Corporation Ltd, which is the least dependent of the big three companies on its transnational parent companies for management. There would be considerable difficulties in bringing a mining corporation acquired in this way fully under state control, given the overwhelming number of expatriates or existing management staff that would be employed, but it might provide a working base for the future.

2 Creating a Namibian Workforce

The present position

Mining companies in Nambia employ whites for almost all skilled and managerial posts, a total of some 3 900 people, and 22% of the mining labour force (Table A1). At Tsumeb all the 500 graduates, and over 85% of the thousand specialists, are white. At Rössing the figures are 78% of skilled workers (grades 9-11) and 97% of officials grade 12 and above. (Appendix Tables A9). In 1982, there was only one black graduate engineer (at CDM), and a metallurgist had also graduated from the CDM bursary scheme. If anything, the proportion of whites has increased over the last two decades: Tsumeb Corporation had 18% in 1957, and 22% today; CDM 18% in 1965, and 24% today.

This record of past discrimination will be met by an incoming independent government. On the one hand Namibians of whatever colour will presumably be welcome provided they are not themselves racist and are prepared to work under the new regime. Skilled people will be particularly important to the continued operation of the mines. On the other hand, an independence government and mine workers themselves will almost certainly expect 'non-racial' to mean that eventually blacks obtain skilled jobs in proportion to their ratio in the Namibian population as a whole (95%). However, there is unlikely to be a serious clash between these two objectives. Many existing staff, with South African or, at higher levels, British passports, may well leave the country. A significant group came to Namibia from colonial Rhodesia. Even today, the turnover of whites is quite high, with 29% leaving Rössing in 1981 and also 29% leaving Tsumeb between 1977 and 1979. A Namibianisation policy, to promote black Namibians, would be essential as well as just.

South Africa has identified, to its own satisfaction, several different ethnic groups among black Namibians. One of these, the so-called 'Coloureds' — is in a special position. Coloureds have had better education, and have reached higher positions on the mines. They constitute 20% of skilled workers at Rössing, for example, compared with only 2% for other blacks. They will require special consideration in any Namibianisation plan. It is popularly believed that a significant number have come to Namibia from South Africa, looking for work, and that they may return at independence. On the other hand, their job status and general economic and political prospects in South Africa would objectively be worse than in Namibia.

Existing training programmes

Mines have always required training schemes, if only to familiarise workers with the particularities of machinery. In the second half of the 1970s, however, the larger Namibian companies began major efforts to expand and widen their programmes. The impetus was three-fold. Economic growth in South Africa led to a growing shortage of white artisans, and at the same time, the independence of Angola and Mozambique and the escalating war in Zimbabwe and Namibia itself forced South Africa to consider a new order in Namibia, in which a black middle class would play a stabilising role. It could also save money. A 1975 Anglo American report on Tsumeb said:[12]

> The number of artisans is high, as no effective use is made of black labour other than as labourers, carriers of tools and vehicle drivers. It is considered that significant economies could be made in the (engineering) department if company policy allowed more effective use to be made of black labour with a consequent reduction in the artisan zone.

By 1979, the Tsumeb Annual Report noted that savings had indeed been made in the smelter by upgrading artisan black workers and reducing the number of white specialists.

CDM began to train semi-skilled employees to specialist level in 1972. Tsumeb built a new training centre in 1975 and followed suit, introducing blacks to tasks previously reserved for whites. In 1978 Tsumeb admitted its first black apprentices; CDM selected apprentices on a non-racial basis from 1979. In 1978 Rössing, then just reaching full production, introduced a 'cadetship' scheme for 18 students, including blacks, to pay for university study and lead on to high level posts at the mine. CDM's 1981 figure of 125 000 man days of training averaged 23 days training for each worker per year; Rössing averaged three times that on its semi-skilled ('Operators') programme. Further data on the big three companies is given in Tables A10 to A12. Training efforts on other mines are almost unreported, and can be assumed minimal. The Chamber of Mines estimated that between R5 and R7m. had been spent on training during 1982.

These programmes are inadequate on at least three counts. First, in the current political position, they arouse cynicism, suspicion and distrust among workers (pp. 35, 42). This will only be changed with a new political dispensation, and strong monitoring. Second, these are not Namibianisation programmes. One of the key problems pinpointed by workers is that even when a worker has been trained to a particular level, he has to wait until a vacancy occurs until he is promoted. There is no explicit policy to employ expatriates on short term contracts, and at the end of a contract to remove an expatriate once a Namibian is qualified to fill his post. The General Manager of Rössing was even reluctant to entertain the notion of 'expatriate': even foreign employees he regarded as committing themselves to the company for life, rather than for a short term contract.

Third, the programmes are insufficient to meet the need. Companies refuse to release even the most generalised projections of

---

12. Christie, n.d.

Table 8

Employment of Citizens — Comparative Overall Performances Since Independence

Country/Mine	Date of Independence	% Expatriates																
Years after Independence		**1**	**2**	**3**	**4**	**5**	**6**	**7**	**8**	**9**	**10**	**11**	**12**	**13**	**14**	**15**	**16**	**17**
Zambia																		
All copper mines	1964	15			10						8		7		5		4	
Botswana																		
All mines	1966	12								13	12	11	11	10	9			(7)

Notes: Figures in brackets are projections.

Sources
Zambia: CISB Mining Year Books. Daniel 1965, p. 198.
Botswana: 1967 Labour Census; 1980 Mines Dept Annual Report; Fifth National Development Plan.

how they expect their training programmes to affect who holds what job in the future. Without published projections, it is impossible to quantify the shortfall. Nevertheless, Fig.5 and Table A11 show that the annual output of trained graduates and apprentices falls far short both of the number who leave each year, and of comparable efforts in Zambia and Botswana.

The major impact of black training has been at the semi-skilled level. Table A9 shows progress at Rössing, where even in the lowest skilled grade still only 7% of the workers were African at the beginning of 1983. A similar programme has been underway at CDM. At Tsumeb, though, the programme to train 'aides' was curtailed in 1981, as a result of 'saturation'.

The overall pace of future Namibianisation

Fig.5 and Table 8 show what has already been achieved, and is projected, on the Zambian copper mines, and on Botswana's principal diamond and copper/nickel mines. In five years the number of expatriates was cut to 10%, and in another five years to between 6% and 8%. By the 16th year, there were only 4% expatriates in Zambia, with perhaps half as many posts again unfilled. Botswana hopes for only 1% (20 expatriates in all) on the Orapa/Letlhakane diamond mines by the 12th year, although in practice there will be more since that figure does not allow for any drop-out of Tswana employees. Tanzania's smaller but more complex Williamson's Diamond Mine had replaced all expatriates by about 16 years after independence.

If Namibia followed the same sort of path, its mines might take longer to reach the 10% level, because they start from a higher proportion of whites, and possibly because colonial education in Namibia is even worse than it was in Zambia and Botswana, so that there are fewer suitably qualified entrants.[13] A rough estimate might be for the number of expatriates plus Namibian whites to fall from 3 900 now to 2 300 (13%) after five years, and perhaps 900 (5%, the proportion of whites in the population) after 15 years. These figures would have to be checked against the numbers of school-leavers available for the mines.

Events could turn out to be very different. Mines could close, or new ones open, altering demand. If independence is delayed, the

---

13. From the point of view of skill, after nine years in Zambia 85% of skilled posts were Zambianised, and 25% of official grade and above: the same percentages applied to Rössing would bring the total number of whites down to 10% in nine years. However, given a lower skill level at Tsumeb and in the light of Botswana and Tanzania diamond experience for CDM, the overall Namibian picture should be better.

Figure 5

Employment of citizens: Botswana compared with Namibia

Base Metals

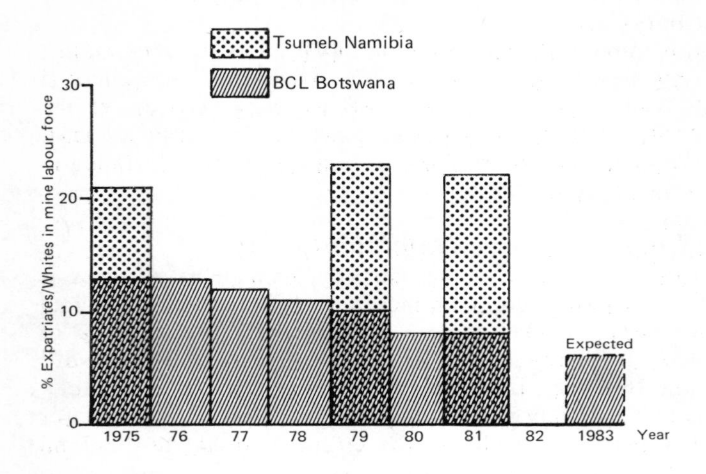

Diamonds

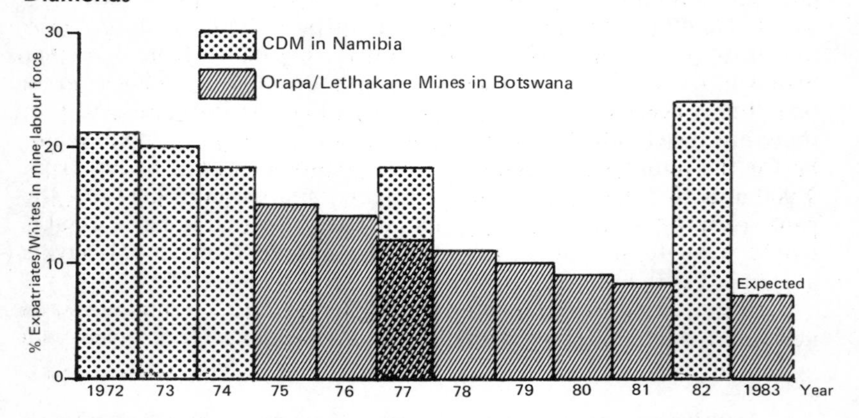

Note: The mines are not directly comparable – BCL is a copper/nickel mine, for example, which smelts but does not refine. Likewise the Botswana figures are for expatriates, and the Namibian for whites. Nevertheless, the difference is striking, and so is the steady effect of training Botswana citizens.

Source: Table 8

training policies of the existing companies are crucial. So are the decisions of Namibian white staff. Government could also influence the rate of Namibianisation by deciding to allocate more, or fewer, qualified school-leavers to the mining industry for training. Whatever the policy, however, a decision to continue the mining industry will imply continued dependence on expatriates in the medium term.

UNIN has estimated that 500 posts could be taken over by black Namibians shortly after independence.[14] There are undoubtedly supervisory posts that can be taken by experienced workmen, and administrative and clerical posts are another obvious area. Rössing has some 250 whites in semi-skilled or part-skilled jobs (81 of them female); over 400 whites at CDM are semi-skilled, mainly women in clerical posts. Some of these workers are, of course, Namibian rather than expatriate; nevertheless, if the employment pattern is the same in other mines, the UNIN figure does look feasible, implying replacing about a third of them.

Namibianisation strategy

The key tool is a Namibianisation plan, showing numbers trained and posts Namibianised over, say, a ten-year period. Such a plan depends on a range of assumptions, principally,

(i) the numbers available to enter training, and the educational standards of these entrants,

(ii) the length of training courses and subsequent on-the-job experience required for each post,

(iii) the number and level of posts to be filled,

(iv) the number of people who drop out.

Since the assumptions are uncertain now and will change over time, there is much to be said for a computerised model on which alternatives can be assessed. Such a model was produced by the Manpower Planning, Training and Zambianisation Unit for the Zambian mining companies.[15] Rössing and perhaps other Namibian mines have begun some work on these lines which may also be useful.

Such a tool should not, however, be mistaken for a strategy. The crucial strategic and political issues show through in the assumptions which the model needs. There are conflicts to be assessed: for example, an expansion of mining will tend to reduce the proportion of senior posts filled by Namibians. Above all, it is essential to question the existing structure of jobs on the mines: simply inserting more blacks into the existing structure is unlikely to achieve fundamental change. One change is already happening: Namibia's present mining

14. UNIN 1978.
15. Blom and Knights 1975.

companies are reducing the proportion of expatriates by breaking down some skilled jobs into component parts which can be undertaken by semi-skilled Namibians. This has advantages and dangers, which are discussed below. It is not, however, the only possible change. Government and miners may have interests in greater worker participation and less hierarchy. Likewise, at the very top, government may wish to see managerial and technical duties delineated so as to allow genuine responsibilities for Namibian managers, who because of apartheid do not have professional mining expertise. All this will change the jobs for which people need to be trained.

If transnational corporations (TNCs) are retained as mine operators (either directly or in joint ventures), government will need mechanisms to prosecute its training policies. A detailed flow of information and regular consultations are essential to successful and realistic planning. The principle of this should be acceptable to companies. In Botswana the diamond mining company produces a five year forecast which is reviewed annually and is used by the Labour Department as the basis for issuing work permits for expatriates. There are precedents for government/TNC mining agreements covering citizenisation in considerable detail, with obligations on the company to train and citizenise by specific target dates: the most relevant example for Namibia is perhaps RTZ's Indonesian agreement.[16]

In some African countries, mining companies have effectively and usefully become training institutes for the wider economy, with trainees leaving mines for other work. An alternative is for the government to set up training schools separate from those of the companies.[17] At least at *non-professional* levels, training courses would have to be based at mines: swift and effective transfer of responsibility demands on-the-job experience and training for specific tasks that may differ from mine to mine.[18] Planned practice is as important as classroom work. Planned practice, in areas of the mine set apart for this, is the most efficient teaching method: the trainees can make mistakes, and be as slow as they want, without affecting production. Training courses can cover very specific subjects, such as explosives handling, drilling, basic sampling methods, aspects of

16. This is examined in Ushewokunze 1981.
17. Botswana's experience, however, was that for non-professional training companies were generally more flexible and swift to respond than state institutions.
18. Information in this paragraph from Lawrence Morris, a mining engineer serving as a CIIR volunteer in Nicaragua.

milling. The length of courses can range from two weeks up to one year at most.

There are precedents elsewhere in the world for *professional* staff to be trained abroad by mining companies. TNCs can also provide on-the-job experience at their mines in other countries. However, in most cases professional training is conducted at universities or mining schools: the link with the TNC is not essential. In any event, an incoming government in Namibia may want to consider carefully where the ultimate loyalty of Namibian professionals should lie: to a transnational corporation or to a developing Namibia. There is much to be said for requiring Namibian professionals to be state employees and appointing them as necessary to privately owned or joint venture mines. In that case, the state could organise their education (perhaps funded through a training levy), with mining companies providing subsequent experience. At the least, professionals trained by the state can be 'bonded' to work in the government service for a certain number of years after qualifying.

3 Labour Organisation and Worker Participation

> Merely taking means of production (machines) out of the hands of their private owners is not enough to ensure their satisfactory socialisation . . . We can speak of socialising only when . . . on the basis of his work each person is fully entitled to consider himself a part owner of the great workbench at which he is working with everyone else.
>
> Pope John Paul II (encyclical *Laborem Exercens*, September 1981)

Independence will release both the energies and the expectations of mine workers. Energies, because workers are part of the independence movement and the atmosphere of antipathy and fear towards the state will have hopefully disappeared. Expectations, because a new government will be in a position to compel changes. In particular, miners will expect genuine training programmes and rapid Namibianisation; an end to racism; a major change in conditions, including wages, housing, health and other services; and significant Namibian presence at the top of the company, so that miners feel both that they are working to benefit Namibia rather than foreigners, and that there are managers sympathetic to their needs.

Whatever form of ownership is adopted for the mines after independence, there is much to be said for swift moves to involve workers and their representatives in the running of the mines. A SWAPO government may regard it as a first step in transition to socialist organisation in the industry. It will, anyway, be important for workers to satisfy themselves that adequate moves are being made in training, promotion and eliminating racism. Particularly if pre-independence management staff are retained, it would be in government's interests to involve workers directly. If planning and monitoring in these areas were, instead, a matter purely between the company and the government, there is a risk both that the government would lack information which workers could provide, and that miners would become disaffected because they are unaware of inevitable practical difficulties that may delay progress. A formal mechanism involving a strong trade union, management and central government representatives would be preferable.

Any such representation first requires organisation. Namibian miners have a history of opposition to exploitation, most solidly shown in the 1971-72 strikes against the migrant labour system. But no open trade unions have been allowed to develop. Underground organisation continues, but most workers have no experience of the usual run of trade union business, much less of participation in taking any level of decisions on mine operations.

There is, therefore, a major task in worker education and organisation. Courses, independent of the companies, will be needed on workers' rights; on methods of organisation; and on the role of mines in the development of a new Namibia. One early initiative could be crash campaigns in literacy and in English language which use learning materials and methods that deal with such issues. The advantage of such campaigns is that they can be undertaken relatively quickly and easily. On the Nicaraguan mines, the number of people who could read and write rose from 30% in 1979 to 85% three years later, as a result of the literacy crusade. Similar results were achieved in the Tanzanian industrial sector over 1973-77 in a programme largely supervised by trade unions and workers' councils.

For a smaller group of workers, more detailed courses could be run at a residential centre, including, for example, understanding company accounts. Establishing such a centre for worker education could be seen as part of a strategy of transition to socialism. But, it would not necessarily spark off opposition from the more enlightened companies, who recognise the advantages of trade unions in creating orderly relations; indeed one possible model for such a centre is Silveira House in Zimbabwe, where worker education is partially funded (with no strings attached), by the Confederation of Zimbabwe Industries as well as the trade union movement.

The exact form of worker organisation will presumably be decided after independence. The standard institutions, according to present SWAPO policy, would be trade unions and worker councils. In Zimbabwe, elected workers committees were set up in each plant immediately after independence, and met with an enthusiastic response. This is quicker and involves less complicated structures than rebuilding a national trade union movement (which however also exists in Zimbabwe).[19]

Namibians will also decide for themselves their forms of management. SWAPO has called for a transition to socialism. If this is adopted, small mine workings using simple techniques might well be organised as co-operatives. Most Namibian mines are large and complex, however, and therefore would almost certainly be state enterprises, or joint ventures. The question of industrial democracy would arise: how far (and how) can workers in a company take decisions, and how far must the general interests of society take precedence? The debate on this is too vast, and much of its application too far into the future, to be reviewed here. Nevertheless, it may be valuable to point to a practical example of transition — the nationalised mining sector in Nicaragua since the 1979 overthrow of the Somoza dictatorship.[20] Conditions in Nicaragua were extreme, in that the foreign companies running the mines had all withdrawn.

During Somoza's time, there were no independent trade unions on the mines, no Nicaraguan professional staff, and very poor conditions. After 1979, with the private companies withdrawn, a new Ministry of Mines was established. At each mine the ministry, known as Condemina, appoints the administrative and technical staff, who are primarily expatriates because of the great shortage of trained Nicaraguans. The workers themselves, however, appoint the men formerly known as foremen.

A great emphasis was put on building a strong union movement, linked to the Sandinista government but with elected representatives at all levels. The trades unions are now the main contact point between workers and Condemina management. At the national level, Condemina has an advisory body which meets monthly and on which sit trade union representatives. At the El Limón mine, each department (there are 32) has its union meeting every fortnight. The departmental meetings feed into an elected 13-member executive council which meets, also fortnightly, with the administration and is party to all decisions about welfare, production and so on.

One underlying conflict emerged at the start in Nicaragua, and will also be present in Namibia. The development of the country requires high mineral revenues which are then applied to support

19. There are elected employee representative committees of various kinds on some Namibian mines. Lacking government backing and with organisation of workers very difficult, they are not the same as the Zimbabwe committees. See footnote 27, p42.
20. Information on Nicaragua from Lawrence Morris, Dick Harding and Jenny Rossiter, and also Nicaraguan Solidarity Campaign 1980.

widespread farming and other development. High production and large surpluses on the mines, however, can conflict with shorter hours of work and higher wages for miners. At El Limón, when the mining company left, the workforce set up a co-operative; they considered that the wealth was theirs rather than belonging to the nation as a whole, and so immediately voted a massive increase in wages for themselves and employed everybody who was then out of work, regardless of whether there were jobs for them. The government on the other hand desperately needed increased production. The result was strikes and in one mine, according to the director of the state mining corporation, a 'tremendous lack of trust between the state and the miners'. A similar clash could occur in Namibia, especially at CDM and Rössing. Profits are seen to be large and yet the government will both be reluctant to create a 'labour aristocracy' by paying wages higher than those on other mines and elsewhere in the economy, and also unwilling to erode the large state revenues these mines can provide (even with substantially higher wages at lower levels of the workforce).[21]

This is a structural conflict which cannot simply be removed. But it is more likely to be overcome if mine workers understand the issues and participate in management. The Nicaraguans believe firmly that there will not be the production needed for the country unless the workers are involved in the running of the industry.[22] The trade union federation in Nicaragua has signed an agreement with the government covering wages and conditions. It has agreed to concentrate on production, considering the needs of the country as a whole in return for the improved 'social wage' they and the majority receive in terms of health and education facilities, guaranteed basic food supplies at fixed prices, and so forth. The federation organises workers' seminars to discuss such issues, distinguishing between immediate demands which (while just), may damage the country, and the longer term interests of workers as a whole.

In Namibia, mine workers are likely to start firm supporters of the independence government. Furthermore, for all the evils of contract labour, it has created unity: as migrants with families left in the rural areas, today's miners do not feel a group apart from other

---

21. It has happened in Botswana that mine workers and a private mining company have agreed on higher wages, but have been opposed by the government. If the company is heavily taxed on profits, an increase in wages can cost the company considerably less than it costs the government in lost taxes.
22. A study of newly-nationalised firms (including metallurgical plants) in Chile before 1973 supports this: despite many difficulties, the higher the level of participation by workers in co-management, the better the economic performance of the enterprise (Barrera, 1981).

Namibians, entitled to special treatment. It is significant that in the late 1970s mineworkers rejected the idea of a separate trade union, in favour of a National Union of (all) Namibian Workers. An organiser reported that mines wanted one voice with 'our brothers who suffer on the farms and in the towns'. This is fertile ground for an understanding between mine workers and government.

4 Abolishing the Migrant Labour System

Miners, their families and Namibians generally have a deep and long-standing hatred of the soul-destroying 'contract' system. 'Contract', and 'the wire' are names for a system that is much wider and deeper than just the employment of single people, without their families. It is an integrated system of overcrowded, poverty-stricken homelands, and a subjugated labour force. A mass meeting of strikers in Ovamboland in January 1972 called it a form of slavery: 'The so-called homelands have become the trading markets where blacks are bought, and in this trade SWANLA (the recruiting organisation) has become richer and richer and the blacks poorer and poorer'.

Miners expect independence to bring 'contract' to an end. This is not simply a matter of law. Indeed, many of the legal provisions forbidding movement and the right to look for work were removed in 1980. But migrant labour still remains. The mines recruit workers normally on short-term contracts,[23] even if expecting workers to renew their contracts at the end of the term. The housing at almost all mines for almost all black workers consists of enclosed, single-person compounds rather than family units.

Workers themselves cannot easily and confidently abandon any rural base they have, even if family housing is available. In town, workers cannot be sure of keeping their jobs. How can they survive if they fall ill, or when they are too old to work? Many do not regard town as a healthy environment in which to bring up children. Farming can provide valuable extra income, to supplement meagre wages, but for very few people does it bring in enough to let them stay in their rural home rather than leave in search of work. Women in particular have little chance of getting a wage-paid job if they go to town. In all these ways, the present structure of Namibia reinforces the contract system.

Abolishing the 'contract' system therefore requires structural changes. Most are outside the scope of this booklet. Indeed, much of the programme of a post-independence Namibia could be described as

23. Rössing is an exception, with permanent contracts, but in unskilled grades only seven days notice is given if workers are dismissed.

'breaking contract'.[24]

On the mines, the immediate physical requirement is for family houses for those that want them. The companies are extremely reluctant to spend the necessary large sums. The companies claim, first, that the workers do not want to move to the often isolated mining town, abandoning their land. This is a hollow argument: the same companies would never dream of insisting that their expatriate white employees leave their families behind because they will not stay for ever on the mines. Workers and families are clear that they want the right to live together, even if not all would wish to exercise it. It is possible that for some months of the year, some of the family may choose to live elsewhere, for example planting and harvesting a crop. But that is no reason to deny them the right to join the rest of the family at other times of the year. Certainly no worker plans to retire at the site of Tanzania's isolated diamond mine; but family housing (and permanent employment) have been provided there for over twenty years.

The companies, particularly Tsumeb and CDM, also argue that it would be madness to build so many houses because the mines will be exhausted in a few years. If true, this would be a more serious argument. However, any such claim needs very careful scrutiny: in this case, it is in the interests of companies to minimise the size of ore reserves they report. Oranjemund (CDM) in particular admits to a fifteen year life, as it has for the last twenty-five years. In practice employment is likely to reduce gradually rather than stop at once. For Tsumeb a different argument applies: the town is in the centre of a prosperous agricultural area, and even if the mine shut down, it is a likely site for alternative jobs to be created: houses would not be wasted.

The type of houses built, the method of building, and who pays for them needs thought by the independent government. If the present companies are retained, then the companies would no doubt be expected to pay; indeed, it might be a condition of their staying in the country. Even so, some of the money spent on housing would be money that would otherwise flow to the government as mineral tax. 838 houses plus associated facilities at the Arandis village for Rössing mine cost R19m. up to the end of 1982. Per house, this is expensive: at the same rate, housing CDM's 4 000-odd African labour force would cost at least R100m. Government may or may not want houses on the mines to be of a much higher standard than the houses that can be provided in rural areas — as houses in Arandis today are, by

24. See, for example, UNIN 1978, Moorsom 1982 and Green and de la Paix 1983.

comparison with today's rural areas. Both within a mining town, and between a mining town and the rest of the country, differences in housing reflect the social structure of the country: high status families get big houses. If there are to be changes in the social structure, then new housing standards are likely to be needed, and old houses may have to be used differently. To solve this problem of giving special privileges to mineworkers compared with other citizens, Botswana tried to charge high economic rents to miners for their houses. Nicaragua after the 1979 revolution went a different way: decent housing was built for miners instead of giving them large wage increases. To do both would have left too little for the mineral tax on which the Nicaraguan government depends.

4.2 Establishing Control

The issues dealt with in the previous section are vital to the future growth of a genuinely Namibian mining sector. The most pressing concern at independence, however, will be to establish control over the present mines, to ensure that they are operated in Namibia's interests. The state will have to play an active role (p.67 above). On the other hand foreign expertise will still be required: the problem is how best to control and use it. This section looks first at the range of possible foreign partners and then at the three main areas of ownership, tax and marketing.

1 Foreign Partners

The range of potential foreign partners in the mining sector is theoretically fairly wide. One option is the present companies, if they are willing to stay. Two obvious advantages are their knowledge of the deposits, and their position in the world market, a particularly compelling consideration in the case of diamonds. Also, a critical factor in the short term, disruption would be minimised. However, the present companies carry the burden of the past: the attitudes of management and skilled labour; the antagonism of the mine labour force (which might be alleviated if a new company was set up as a joint venture between government and the old company); and defiance of international law, especially for mines established since the mandate was revoked in 1966 (Rössing, Oamites and Otjihase, principally).

It is possible that outsiders will try to include in the Namibian independence settlement provisions which wipe the slate clean and ignore the past actions of the present mining companies. Any such provisions, especially for the post-1966 mines, would ignore both the decisions of the International Court of Justice, and the history of the companies. At the very least, the past profits of a company must be taken into account when the government calculates what level of future profits would be reasonable for the company, or how much compensation it should get for assets acquired by the state.[25]

Alternatives

There are alternatives to the present mining companies. If the mineral market is not too depressed and if the government is seen as a reliable partner, other Western companies may be interested. These conditions might hold for diamonds. Significantly, given the international legal position of Rössing, the same conditions may also apply to uranium, for rather special reasons, including customer concern at safeguarding their source of supply. There are other Western companies that might be interested. The lack of new investment in post-independence Zimbabwe is, however, a warning of the general attitudes of companies: especially in a depressed world mineral market, it cannot be assumed that companies will be prepared to come in unless the prospects of profit are very clear.[26]

A further possibility is enterprises from socialist countries. The USSR has relevant mining experience, including extensive diamond mines and a diamond cutting industry. However, the mines are of a different type, and it is notable that Angola has turned to a De Beers-affiliated company for assistance with its diamond mines.[27] A number of Eastern European state mining agencies have entered into agreements with Third World governments.[28] They require much the same terms as Western transnational corporations: any foreign partner, whatever its origins, needs a minimum return on its investment.

A different approach would be to recruit individual foreigners to fill all the posts. This is the route followed by Nicaragua, after lengthy

---

25. One possible negotiating strategy for Namibia, if existing companies were to be retained, is to seek a major equity holding in existing mines in return for confirmation and extension of mining rights. The size of government holding could take into account the size of previous profits. It is common, in new projects, for governments to receive shares in return for mining rights and geological data.
26. A more hopeful comparison is Angola, where new oil companies have been interested in investing.
27. The USSR also markets a large proportion of its diamonds through De Beers channels. It would of course be a competitor of Namibia in diamond sales, which are predominantly in the West.
28. An often-quoted example is the agreement between Kenya and Romania's Geomin for proving a lead-zinc-silver deposit. The government was to pay for its 51% of equity, and Geomin for its 49%. Both sides had three directors in the joint company, giving an effective veto on significant decisions (Zorn 1978).

negotiations for agreements with foreign enterprises had failed. However, even with smaller and technically less complex mines than in Namibia, Nicaragua had three years of major disruption because of difficulties of recruiting technical staff. It would probably not be initially practicable for Namibia to recruit enough individual staff to maintain the mining sector near its present levels. Nicaragua's experience does suggest that, if it became necessary, sufficient technicians might be found to keep crucial mines running. But given the time and expense required to find suitable people, individual recruitment is unlikely to be the best method of obtaining foreign skills.

2 Goals

In negotiating with these potential partners, the new government is likely to be concerned with three main areas. First, that the mines are run productively and profitably in accordance with good mining practice. Second, that the mines follow government policy, particularly towards disclosure of information, training, organisation of work, migrant labour, health and safety, and the abolition of racial discrimination. Third, government will want to ensure that foreigners do not take excessive profits, whilst recognising that foreign enterprises will have minimum requirements of their own which must be met if there is to be successful co-operation.

One of the most important of these requirements of foreign partners is that the arrangements negotiated will be stable. This does not mean that relationships between Namibia and foreign partners need remain static: contracts can provide for change, for example the rapid promotion of citizens, and can include procedures for unexpected outcomes — such as additional taxation if profits are higher than predicted. Review of contracts may be necessary, after several years in operation. However, in general the more stability can be provided, the more confident partners are likely to be.[29]

3 Ownership and Management Options

These goals have to be applied in a number of areas, and most obviously in the area of who is to own and who manage the mines. It can be assumed that government will want to take at least some stake in the principal companies. Nevertheless, an important lesson of the last two decades in Africa is that greater state ownership of mines does not by itself meet government objectives of the kind outlined above. Whatever the level of state shareholding, foreign expertise will still be required, and will still have to be controlled.

---

29. Examined in Brown and Faber 1980.

Mining companies themselves often welcome limited state participation, reasoning that government will then protect, rather than attack, the company. The chairman of De Beers, for example, said in 1974,

> No government likes its basic industries to be entirely foreign-owned, and yet in many developing countries individual members of the public either do not have the resources to invest in industry or for ideological reasons are prevented from doing so. The only alternative in such cases to full foreign ownership is for government to take a direct interest. In these circumstances we willingly accept a partnership between the government as owners of the mineral rights and private companies that can provide the necessary financial resources and technical knowhow.[30]

Perhaps, significantly, the 1982 President's report to the Namibian Chamber of Mines referred to the dangers of state control and state management, not of partial state ownership.

There are, in principle, a range of alternatives, from full nationalisation of the mines to a small minority shareholding by the state. Examination of these alternatives for Namibia's various mines (for the answer may not be the same for all) has already begun, and to avoid duplication they are only briefly summarised here.[31]

(a) *Full state ownership* theoretically allows complete state control. However, either foreign staff have to be hired individually on the world market, which can be difficult and expensive, or a foreign enterprise has to be given a management contract. Contracts need monitoring, all the more so since the foreign enterprise has no capital of its own at risk if the mine fails to be profitable. Nationalisation is likely to antagonise the existing operating company, which would create difficulties if government wishes to retain it to manage the mine.[32] Nationalisation also maximises the expense if compensation has to be paid. At worst, nationalisation with compensation, management by the former owner, and sales controlled by the former owner can add up to a package which raises costs, expatriate employment and the profits of the former owner but does little or nothing for overall national control, surplus mobilisation or development: e.g. Gecamines in Zaire.

(b) A common arrangement in Africa is for the government to hold

30. Statement to shareholders at 1974 AGM of Anglo American Corporation, quoted in Brown and Faber 1977.
31. Zorn 1978, Aulakh and Asombang 1983 for more detail.
32. In Chile before 1973, TNCs whose Chilean mines had been nationalised even managed in retaliation to block sales. In general, however, courts will not act against minerals produced after nationalisation.

51% of the equity of a mine. Theoretically state control is retained, but the foreign company has a stake in the profitability of the mine. In practice, foreign partners generally insist on veto powers on key decisions, especially those that involve investment of their funds. In the Botswana diamond arrangements, the effective power of both sides is explicitly recognised in the formula that company and government *each hold 50% of equity*.

(c) A *minority government holding*, of whatever size, gives a direct share in profits and allows the government information through a seat on the board. However, crucial details may nevertheless be concealed, since many do not go to the board in the normal run of business. The extent of influence on mine policy depends on the terms of the agreement.

Management

Management is closely related to ownership. Foreign companies often wish to retain as strong a hold as possible on the management of a mine, whatever the ownership arrangements: they believe that they are the best judges of the road to profits. To the extent that they are efficient, such an arrangement may appear attractive to governments. Nevertheless, two problems arise. First, control of management gives rise to control of information; the government then may not be told about key decisions, or may not be presented with the full range of options (see Appendix 4 for an example of a technical issue in which the government has an interest). Alternatively, management could siphon profits abroad through transfer pricing of some kind (see next section). Certainly, government suspicions may be aroused, whether justified or not. Second, management solely by the foreign company can create a particular management style, and a way of organising work which goes against goals of increasing participation by the workforce. To meet these problems, there is no substitute for direct Namibian participation in management. Immediately, there would be a severe shortage of skilled Namibians; nevertheless, the involvement of independent government officials can be effective, even if they are foreigners.[33]

4 Government Revenue

The share of proceeds going to the state from mineral revenue is obviously related to ownership of the mines, but it is not the same. Even if mines were purely private enterprises, governments could tax a high proportion of profits. For example, although the Botswana government owns only 50% of the shares in the diamond company, it

33. Ushewokunze 1981.

is said to receive 75% of the profits when tax is taken into account. Indeed, it would still receive perhaps 50% if it owned no shares at all. Conversely, government could receive nothing at all from a fully nationalised mine, if fees for a foreign management contract and payments on foreign loans left the mine with no profits. The amount of income that an independent Namibian government might obtain is considered in Section 5.2 below; here we are concerned with methods.

The merits of different kinds of taxes and other payments to government have been extensively discussed, and need not be re-examined here.[34] One approach starts from the basic principle that a foreign company requires a 'reasonable return' on any capital invested: there is likely to be some minimum expected return that it would need to attract it to Namibia. In principle, the government can then take the rest of the profits over and above this minimum. The government share will be more or less depending on how profitable (i.e. how low cost) the mine is: we are back to the theory of rent.[35]

This basic principle needs to be modified. In the first place, there is considerable uncertainty, especially with new projects, about what profits will be in the future. At the same time, both governments and foreign partners have an interest in as much stability in their contract arrangements as possible. Governments often want to ensure the country receives some tax revenue even if the mine is running at a loss. This is most simply done by imposing a tax (called a royalty) on the total value of minerals sold.[36] Conversely, if profits turn out higher than expected, the state does not necessarily want to take all the surplus, over and above the company's 'reasonable return': instead it may give the mine a sufficient share in extra profits to encourage efficiency. 'Additional Profits Tax' is one way of achieving this.

In Namibia, a further issue arises. Most, if not all, of the present mining companies have already had a more than reasonable return on their investment. This can be taken into account in any consideration of compensation, and it should also mean that such companies would accept a considerably lower return in the future. They have, in practice, recovered their sunk capital so that the risk of major losses is negligible. However, there is still some minimum average annual cash flow required to make it worth their while to allocate continued

34. E.g. Ushewokunze 1981, Aulakh and Asombang 1983, and Brown and Faber 1977 and 1980.
35. A practical implication is that, when negotiating, all financial arrangements should be dealt with as a single package. There are many other issues, such as how tax payments or losses in Namibia may affect a foreign company's tax payments in its home country.
36. In 1983, reportedly acting on IMF advice, Zambia introduced just such a royalty on its copper exports at a time when mine profits were very low or non-existent.

attention and personnel to Namibian operations, instead of shifting to new ones — a minimum which is higher for restructuring or opening new mines than for day to day operation of existing ones.

Monitoring transfer pricing

Whatever the tax regime, the independent government will wish to monitor financial arrangements, to ensure fair dealing. It is possible for transnational corporations to overcharge a subsidiary in Namibia for the services of their head office. The Anglo American report on Tsumeb Corporation comments that the amount charged by the United States refinery for refining Tsumeb's blister copper 'seems to be excessive' — the refinery was owned by AMAX, one of the main shareholders in Tsumeb. Alternatively, the local subsidiary can be underpaid for goods supplied to another part of a transnational corporation.[37] Such 'transfer pricing', as it is called, is relatively well-known internationally. Methods of monitoring have been developed, by African countries and by the United States Treasury Department and Internal Revenue Service. In practice, none of these agencies have been strikingly successful in their efforts.[38] It is an area where international and regional co-operation (for example, with the framework of the Southern African Development Co-ordination Conference), may be useful and cost-effective. Direct public sector intervention in export marketing — as practised by Zambia and being instituted by Zimbabwe in respect of metals — may be successful in preventing transfer pricing on sales of minerals, but cannot meet the problem of purchase of goods and services by the mines. There can be no substitute for detailed scrutiny of the mining enterprises themselves.

5 Marketing

Namibia's interests in selling minerals are obvious. First, that the minerals be sold to buyers offering the best prices, along with a reasonable prospect of longer term stability; and second, that the contracts are fair and fairly implemented, without 'transfer pricing'. The first requires knowledge of the available buyers and of trends in world markets, and the ability to react quickly to changes — in short, knowledge and contacts. The second requires careful scrutiny. A third interest of Namibia, as with all mineral-producing governments, may be to impose conditions on sales: for example, uranium producers

37. Eric Lang asserted, though without quoting his source, that this was responsible for R100m. lost taxes in Namibia in 1982 (WA 25 May 1983). The Thirion Commission has also made trenchant comments on the present lack of scrutiny of mineral taxes (WA 26 Jul 1983).
38. The issues are discussed in Robin Murray 1981.

such as Canada will sell uranium only to countries that give assurances that it will be used for peaceful purposes. In practice, except for limits on uranium, Namibia's restrictions may be few: SWAPO has indicated that it would want to sell minerals on economic rather than political criteria. Uranium supplies for European power plants would be maintained so long as a fair price was paid.[39]

It is clear that the state must play a role. Mining companies argue that they have the necessary contacts and experience and can be trusted to seek out the most profitable markets because it is in their interests to do so. But this is not always the case — they may wish to pass the minerals to another part of their own transnational company.[40] In any event, they cannot be relied upon to prevent transfer pricing. The companies themselves should recognise that their long-term security in Namibia should be greater if the state can satisfy itself that it is not being cheated.

Given these objectives, there are a range of new institutions that could be set up. There are very different markets for particular minerals, so again the answer may not be the same in each case.

Historically, there have been four basic methods by which countries have tried to protect their interests:

(a) *Joint-venture companies:* Sales are left in the hands of the various mining companies, but the state has a share-holding in the company and so a say (which varies according to the agreement) in marketing. This is the practice in Botswana, for example, with marketing arrangements included in negotiations for the mining lease. The Botswana Diamond Valuing Company (BDVC) is a 50/50 joint venture between the state and De Beers, with government's interests protected by an independent government valuer. The sorting operation, handling 9m. carats a year, half the total De Beers group production, is carried out in Botswana and is being rapidly taken over by Botswana nationals.

(b) *Legal restrictions:* In 1901 the Australian government took powers to control exports, and since then has at various times forbidden the export of iron ore and coal at prices it thought were too low, and so forced re-negotiation. Such legal provisions are certainly essential if production is to be held back by common agreement with other producing countries (in order to force the world price up), for example, in CIPEC for copper producers.

---

39. The nature of uranium and of global geopolitics adds the consideration that sales be transparent and to reputable civilian users or their suppliers.
40. There is an increasing tendency for buyers, especially electricity boards using uranium, to lend money for the original investment in the mine in order to secure supply.

But on their own they are unwieldy, with no automatic way of making small adjustments, or of monitoring prices.

(c) *Compulsory state marketing:* A state marketing corporation is set up, which buys all minerals from the mining companies and sells them itself. The Zambian Metal Marketing Corporation MEMACO, for example, sells all Zambia's exported copper and cobalt. It has offices in London (since the London Metal Exchange is the key market, which sets the price at which most copper is traded), and contacts around the world. Compulsory state marketing is common among oil producers, such as Saudi Arabia. Zimbabwe's major move in the minerals sector has been to establish a similar body. Tanzania has introduced a diamond valuation unit independent of De Beers and the Central Selling Organisation. The unit ensures that the prices paid by CSO for Tanzania's output are, in fact as well as in form, the final CSO selling price less the negotiated discount. Based in London, this office was initially staffed largely by independently hired expatriates but is now basically Tanzanianised.

Companies dislike state marketing because they fear inefficiency through inexperience and lack of contacts. Nevertheless, in this respect at least, companies have an interest in helping a state marketing corporation to succeed. From government's point of view the main problem, apart from the opposition of companies, is the need to allocate skilled people to the corporation, even for routine jobs where there would be little chance of cheating if the corporation were in private hands.

(d) *State marketing option:* A smaller state marketing body is established, with the legal right to market some or all mineral production. It normally sells only a small proportion (to keep its hand in the market), but intervenes to market more if it believes the mining companies are not getting a good enough price. This has the advantage of requiring a smaller staff than compulsory marketing, and of providing an incentive for companies to play straight. However, the potential for conflict is large, and the uncertainty of this arrangement would be disliked by the companies.

These four methods, or three if legal provision is regarded as inadequate by itself, have their own advantages and disadvantages, but they are best looked at with reference to specific minerals.

Base metals

Copper and lead, the two principal Namibian base metals, are sold on short-term or annual contracts, with prices fixed in relation to those on the London Metal Exchange at some specified date or period. At present, these contracts are made by TCL. The existence of a clear market price considerably simplifies the problem of monitoring transfer prices, although it does not solve it: refining costs in particular need to be checked, and also other charges and discounts, as well as the exact nature of the link to the LME price.[41]

MEMACO of Zambia has proved the feasibility of a state marketing corporation in base metals, and this would also have the advantage of breaking the chain of in-company transactions. Namibians would be selling their own minerals. The question for an independent Namibia is whether it is sufficiently high priority to divert enough scarce, skilled labour away from other tasks. Foreign expertise can, indeed would have to, be hired, but it too would require supervision. The decision may depend on strategy elsewhere. If, for example, the state took a major part in TCL, then there would be a strong argument for concentrating staff there, rather than establishing a separate body to market base metals primarily from TCL.

An early problem may well be finding alternative markets for tin concentrate and zinc from Uis and Rosh Pinah at present sent to South Africa. The economic position of these mines is unclear due to a lack of published material. ISCOR's ownership could mean allowing either below-market prices to be paid for the minerals, or a high cost mine to be kept running in order to secure a strategic supply for South Africa. Formally, as open markets exist for both products and the Namibian supply is a trivial portion of the global total, there would probably be no great difficulty in redirecting sales.

Uranium

Uranium is sold on long term contracts (three to seven years) as well as through one off 'spot' transactions. There is no clear market reference price. The NUEXCO spot market price is often quoted, but very little of the world's uranium is actually sold on the 'spot' market and 'spot' sales by producers do not necessarily occur at the NUEXCO price. Contract prices can be very different from the NUEXCO one, above in years of surplus supply (as 1983), but well below in years of shortage (1977-78).

Negotiation of contracts is therefore extremely important. A small number of large TNCs dominate the market. At least during the early 1970s, they formed a cartel (including government and company participants), controlling the price.[42] Companies rather than countries

---

41. See K.M. Lamaswala — 'The Pricing of Unwrought Copper in Relation to Transfer Pricing' in Robin Murray 1981.
42. See Radetzki 1981. In the mid-1970s some six units controlled two-thirds of world production (p. 111). Since then, however, with the build-up of Australian production and cuts in demand, over-supply has brought a rapid fall in prices.

make the contracts. Such a secretive and integrated world market increases the need for the state to participate in agreements and check their working. But it also make it more difficult: information and contacts are not as easy to find as with base metals. This is particularly so when the world market is over-supplied, and there are not many buyers looking for contracts.

After independence, a new Namibian government will seek legal contracts to replace the present illegal contracts. SWAPO has said it would negotiate them on a straightforward economic basis. In the present depressed market, it will not be easy to obtain a good price (see p.63 above).

Namibia's principal advantages in negotiations with potential uranium buyers are, first, an operating mine with a high profits record; second, that its existing customers (especially German) wish to ensure continuity of supply; and, third, that some potential buyers want to diversify their supplies, so that disruption does not put their operations at risk. The French in particular have invested widely in Africa for this reason

In this kind of market, it seems inevitable that the negotiation of sales contracts will go hand in hand with the negotiation of an operating agreement for Rössing with a foreign technical and investment partner: the one cannot go ahead without the other. Government would need a heavy input of skill, knowledge and negotiating ability over a relatively short period (perhaps two years) whilst the contracts are agreed. The requirements are very different from marketing copper, with its annual contracts and daily price changes. This might suggest a joint venture negotiation, rather than a uranium division for a state marketing corporation. However, the difference between the two should not be exaggerated. A joint venture arrangement would still require a state role, while the uranium division of a state marketing corporation would have to work closely with the operating company in negotiating a contract.

Diamonds

Diamonds are a third, different category.[43] The diamond market is highly organised. The world price of a diamond is artificial, several times greater than the cost of mining it. It is maintained at that level by (i) controlled production and buffer stocks, (ii) the confidence given to buyers by the uninterrupted steady rise in the price of diamonds since the 1930s and (iii) advertising. Limits on production are in the long run due to the limited number of diamond deposits

---

43. Despite the publicity given to Epstein 1982, Timothy Green 1981 is a more accurate popular account of the world diamond business.

discovered, but De Beers also deliberately cuts back production at its mines when demand is weak. The rising diamond price is partly due to the long economic boom in the West since the second world war. But market management is also responsible: De Beers' Central Selling Organisation, which directly controls over 80% of world gem sales, varies the volume of diamonds sold, and orders producers to stockpile when prices threaten to fall. Advertising is solely the responsibility of the CSO. Their proudest achievement is to have raised the number of Japanese women who buy diamond engagement rings from 5% in 1967 to over 62% in 1981. By 1983, the CSO was spending $72m. worldwide on advertising, to fight the present major recession.

There is little doubt that the present diamond market position is serious, although 1983 is showing some improvement. The underlying foundations of supply and demand are shifting. On the one hand, more diamond deposits are being discovered: De Beers' own production rose from 7m. carats in 1965 to a planned 19m. in 1983. Total world sales in 1981 were around 47m. carats, but the new Ashton mine in Australia could produce another 22½m. annually. The market will not be swamped: only 2-3m. carats of Ashton's production will be gem (as opposed to industrial) diamonds, and even the gems are mostly in the lower quality grades. Ashton's gems nevertheless represent a 20% increase in world gem sales, and it is not the only new find.

Meanwhile, the post-war economic boom in the West has ended. So far demand for diamonds has, in fact, held up well: 1981 and 1982 jewellery sales were, in real terms, only slightly below the 1980 record peak. But can demand continue to increase as it did during the 1970s, at an average of 15% per year?[44] The answer is not self-evident. Demand for small diamonds has responded well in the past to advertising and promotion, and there are new markets that can be developed. Demand for larger stones will normally pick up with an improvement in the world economy. Even in a crisis, diamonds historically retain their worth and can be moved rapidly and discreetly in an emergency — 'Diamonds are a refugee's best friend'. However, De Beers have accumulated a large diamond stockpile, valued at the end of 1982 over R1 800m. at cost price — the sales value will be considerably more. De Beers will probably hold back these stocks whilst reducing production to suit demand, and reckoning on price increases. Stock can then be gradually released, realising additional profits. This has succeeded in the past and, despite the fears of some observers, looks likely to do so through the present recession. But the

44. Figures from De Beers Annual Report 1981. The overall issue was discussed in FT 4 June 1982.

financing costs of such an operation are very heavy, and a prolonged world slump combined with additional diamond mines would threaten the prosperity of the industry.

Especially with a weaker market, the prosperity of diamond producers depends on a continued restriction of supply and mangement of the market. The accepted manager at the moment is De Beers, through the Central Selling Organisation: the USSR, Angola and the new Australian mining company all want to sell the bulk of their gem output through the CSO. A number of smaller producers sell outside. The biggest was Zaire, which left the CSO in 1982 in protest at its unreasonable charges. After leaving, Zaire was able to sell its total (falling) production, almost all low grade 'boart' for industrial use, whereas the CSO would have imposed a quota limiting sales because of the world market crisis. This seemed an advantage but in practice the price of boart collapsed, according to some reports because the CSO was dumping boart onto the world market. At the same time Zaire was worried about competition from the new Australian mines, and in 1983 Zaire rejoined the CSO.[45] In any case, it would clearly be disastrous for large gem producers like Namibia if every country sold its full production independently into a buyer's market. Quotas are required to hold up the price.

But need world sales be co-ordinated by the CSO? Alternatively, if it be granted that the CSO has specialised expertise and contacts in the trade,[46] need the CSO be controlled by a South African private company (De Beers)? It is certainly against Namibia's interests to be used as one of the principal taps to regulate the world market. Independent states, such as Botswana, Angola and Tanzania, have done better than CDM, both at reducing commissions and discounts, and at securing relatively low output cuts when the market weakened.

An even distribution of quotas and cutbacks, and more direct producer control over the CSO, would appear logical goals. A share in its trading profits would also be welcome. The diamond trading companies in the De Beers complex chain are allowed to make gross profits of 6 to 12% of standard selling price.[47] Issues also arise over

---

45. FT 26 Jan. 1982; RDM 25 Jan. 1982, 5 Apr and 25 Apr 1983; *Guardian* 9 March 1983. The price of boart would probably have had to fall to meet competition from synthetic diamonds.
46. Which it has: the CSO 'assortment' does exist and can be checked, whereas other firms have no such classification system. The best diamond markets are CSO clients. However, CSO is far from perfect: it failed to release enough diamonds to meet, or to raise prices rapidly enough to halt, the speculation of 1977-78, which led to the present collapse in prices.
47. According to RSA 1973. Murray 1978 confirms this with the report of an independent consultant, but De Beers disputed the figures.

the financing of, and the profits from, stockpiling. Botswana, which with the opening of the Jwaneng mine will equal or overtake South Africa in volume of carats mined, has shown signs of wanting more information and a genuine voice in marketing decisions. Namibia's production is much smaller in total carats (1½m. compared with 9m. in Botswana); but because 95% of Namibia's diamonds are gem grade, it is critical in the world market. An agreed marketing stance between Botswana, Namibia and Angola (and possibly others) would be extremely difficult for the CSO to resist — but perhaps hard to achieve.[48]

With a central world selling organisation like CSO, Namibia's main needs, as far as its own production is concerned, is for sorters to sort the diamonds, and then for an experienced valuer (initially expatriate) to check that CSO offers a fair price. On top of that, Namibia needs to build up expertise in the more long term strategic assessment of market conditions, in order to negotiate with the CSO and help decide its policy. Both are areas where co-operation with Botswana, Tanzania and Angola might be effective.

It has been pointed out that one consequence of the present selling arrangements is that CDM's stocks of diamonds are both large (perhaps R340m. at full selling price) and held outside Namibia.[49] De Beers could therefore hold these stocks as a hostage against any unfriendly moves by a future Namibian government. An independent Namibia would presumably seek to move into line with other diamond producers in holding any stockpile inside the country.

4.3 Priorities in the Aftermath of Independence

This chapter has considered a wide range of issues. At independence, however, the inheriting government will be faced with everything at once, and will have to set priorities. The date of independence is uncertain. It could come after a bloody victory, with immense destruction, or in a smooth transition, hedged about with limits on the freedom of action of the new government, or anywhere between these extremes. Speculation is, therefore, unsafe. However, it is safe to say that if independence comes in the mid-1980s, the past will impose considerable constraints on policy.

First, revenue would be dependent on the continued operation of the mines, or at least of Rössing and Oranjemund. That would be

48. The implication is that those who have done better in the past would have a reduced share of total diamond output when the market was weak. This, and the problems of cartels in other commodities, may be barriers to moves to assert greater producer control over the CSO.
49. Information from Martyn Marriott.

most smoothly ensured by the continued involvement of transnational corporations, although not necessarily the same ones as at present. This might incline the government to minimise the changes it introduces, in order to retain the companies. Certainly TNCs have considerable potential for disrupting the economy if they are dissatisfied. They could delay investment in new machinery and allow the old machines to wear out; halt prospecting and remove records and blueprints; borrow money from foreign banks and governments rather than commit their own funds, and so increase the number of parties interested in the status quo. At worst, they could sabotage installations as they left.

However, this need not rule out the possibility of major policy and structural changes. The experience of independent Zimbabwe is that TNCs are extremely cautious about new investment, however favourable the government is towards them. A hesitant policy may not attract new investment anyway. Experience elsewhere indicates that what matters to TNCs is that they know where they stand, that the government's position is clear for as far into the future as possible, and that it provides the possibility of some profit to the TNC. In general, companies are happier to invest under a predictable regime, however radical, than in a country where policy is unclear. Oil companies in Angola, for example, have expressed satisfaction, despite the Marxist orientation of the government, with its businesslike attitude.

A second major constraint is the risk of disruption by South Africa. This need not be overtly military since the economy, including mining, is very dependent on South Africa, and therefore open to destabilisation. A major response to this would be international pressure applied on Namibia's behalf. It is extremely important that the international community understand and support the changes required to redress injustice and create a development strategy which takes into account the needs of all Namibians.

The over-riding constraint, however, is likely to be the shortage of skilled Namibians. The mining sector as a whole is unlikely to receive more than its fair share of the available labour and expertise: mining may be a vital source of revenue, but other sectors are more directly concerned with enabling the population to meet their basic needs. Within the mining sector, fundamental change is only likely if thought and personnel are focused on crucial issues. The government has to decide what these are, and, more difficult, to decide what has to be neglected as lower priority, even if desirable.

The most urgent problem will be to arrange working arrangements for the mines. The key concerns are CDM, Rössing and TCL, with Rössing the most pressing because of its past history.

Inevitably this will take up most of the time of available senior officials in the immediate aftermath of independence.[50] Even if change was to be minimised, it would be necessary to negotiate in order to clarify the new basis of operation so that production was kept up.[51]

To formulate more detailed suggestions would be to trespass on the areas where Namibians will make their own decisions. If, however, it was agreed that the crucial long term issue is to lay the foundations for a genuinely Namibian mining sector, then perhaps there is an argument for short term moves to centre on:

(a) major training programmes, committing the mining companies, but also obtaining foreign government-to-government assistance;
(b) creating a national mining service, including one active, operating state mining corporation as well as a capacity for monitoring production, sales and transfer pricing;
(c) joint venture arrangements for the principal mines, with adequate tax arrangements, and with foreign partners providing capital and skilled people for most of the routine work (to minimise both disruption to revenue and demands on government staff) — but with provision for increased state involvement over time as Namibian staff and experience are built up;
(d) building up a strong trades union so as to involve miners themselves, as well as the state, in monitoring and in management;
(e) abolishing migrant labour;
(f) creating an independent diamond sorting and valuation capacity.

Perhaps the most fruitful other initiative in the relatively short term would be to co-operate with other diamond producers so as to

---

50. The likely procedure will be to negotiate mining agreements with foreign partners. However, detailed negotiations are extremely time-consuming. They also occupy a large proportion of the time of very senior government officers, time which will be precious just after independence. Negotiations for the Oktedi mine in Papua New Guinea, even after tax had been agreed, took up half the time of four officers at deputy permanent secretary level for two years, together with over 18 man months of external consultants' time (at a total cost of approximately £1m.) It has therefore been suggested that, where feasible, a standardised arrangement be devised which can be applied to most of the smaller mines — though separate negotiations are inevitable for diamonds, probably for Tsumeb Corporation, and for the post-1966 mines.
51. Negotiations with mining companies were not, in fact, the first priority in independent Zimbabwe — which concentrated instead on gaining control of mineral marketing. However, this is a much less easy option for Namibia, because of its history, the mines being more central to the economy, and because the Namibian mining sector — unlike the Zimbabwean — is dominated by two very large and four to six moderately large mines.

gain greater control over diamond marketing.

Such a package is certainly not the only option. However, it would provide a base for the future, and if clearly outlined would probably be accepted, albeit under protest, by mining companies as a pragmatic approach designed to maintain continuity but also to respect the political interests of an independent government of Namibia.

5 Mining in Overall Development Strategy

The past history of Namibia, and the terms under which independence is eventually won, will constrain the options of the new government. Nevertheless, the government will have goals, and it is impossible to analyse the potential role of mining without making some assumptions about the overall strategy of the new state. Given the popular support for SWAPO, its policies need careful examination.

Two of the basic objectives of SWAPO's constitution are 'To ensure that the people's government exercises effective control over the means of production and distribution and pursues a policy which facilitates the way to social ownership of all the resources of the country', and 'To work towards the creation of a non-exploitative and non-oppressive classless society'. The 1976 Political Programme sets this as a prime goal: 'Economic reconstruction in a free democratic and united Namibia will have as its motive force the establishment of a classless society. Social justice and progress for all is the governing idea behind every SWAPO policy decision'. A major part of the Political Programme is taken up with the need for democracy — for 'involving the whole population in active discussion' and for SWAPO leaders to 'learn about their true aspirations, their doubts and their sense of possibilities'. The Programme goes on to propose specific measures for creating an integrated national economy, with particular emphasis on creating a processing industry and on changing agriculture to meet the food needs of the nation.

It is significant that there is no specific mention of the future role of mining in this basic strategy document, other than the desire to process output. This omission recognises the difficulties posed by mining. Mining provides vital revenue, and it is clear from other SWAPO statements that the government would continue to encourage it. But a mining sector poses considerable problems for a small country attempting to meet the basic needs of the people.

The advantage of a mining sector is obvious: a large supply of foreign exchange allows the government room to manoeuvre. Funds are available both for development projects and for expanding government services. Diamond revenue lies behind Botswana's fast growth rate and improved education, health and infrastructure; oil revenue has been essential to the defence of Angola against constant attack. It is therefore likely that the independence government in Namibia would regard the primary role of mining as to generate income for development. Mines can, in favourable conditions, be geese for laying golden eggs.

Geese are not easy animals to keep, however. They are cantankerous and argumentative; they hiss and snap, and furthermore have a tendency to gang up. In addition, it is not easy to use minerals directly for widespread development benefiting the people as a whole. A mine anywhere tends to be hierarchically organised and, particularly where expensive expatriates have to be hired, social divisions and income differentials are marked. This will be all the more true at independence in Namibia, given the inheritance of present mine organisation, mine towns and migrant labour. It is not an easy place to begin creating the 'classless society' which SWAPO declares to be its long-term goal.

Mining is also not a satisfactory sector for creating many new jobs. It employs very few people for every dollar of capital invested, or every rand of income received. In 1977, although 33% of the GDP came from mines, only perhaps 5 per cent of the labour force were miners. The new Botswana diamond mine at Jwaneng cost R280m. to bring into operation, but employs only 1 300 workers. Jobs can be created more cheaply elsewhere in the economy. The only exception would be very small mines, which the new government could consider encouraging, with processing where necessary at the large mines.[1]

5.1 Problems of a Mineral Dependent Economy

An independence government will wish to tax the mines and spend the tax on wider development. As many oil producers found, this too is not simple.[2] Large engineering, urban and infrastructural projects are easy. It is much more difficult to produce more goods and to create jobs, especially in rural development, that would directly reach the poor. Agricultural projects tend to require technicians who are not

1. In Rehoboth, for example, the Geological Survey has identified a number of potential small deposits, for gold, barite, kyanite, limestone and Iceland spar. (ENOK 1981.)
2. The issues are examined in Seers 1978 and Lewis 1980.

only skilled but also knowledgeable about local conditions and committed to government's goals. Because of colonial policy, skilled Namibians are few, and suitable expatriates will be hard to find. The temptation is to concentrate on what is easy. The risk is to become increasingly dependent on minerals, while the interests of people in the rural areas are neglected. In its worst extremes, mineral income has achieved no more than to reinforce a class of well-paid bureaucrats distributing favours to their friends, as already seen in Namibia in the proliferation of 'ethnic administrations' during the DTA period.

A further tendency of mineral economies, with plentiful foreign exchange, is to import well-known foreign goods rather than to produce them at home. Pressures on the exchange rate and on industrial wages and other costs also operate so that, as mineral exports rise, imports become cheaper relative to locally produced goods. The attractiveness of non-mineral exports (such as cash crops or beef, which may employ more people) also falls. In both ways, local producers are discouraged.

The economy may nevertheless boom, if mineral production and prices stay high. However, enormous problems can be created when prices fall, or when mines are exhausted. This can be illustrated with reference to what is likely to be an important policy concern of an independent government in Namibia, the expansion of social services such as schools or health clinics. Expansion will have widespread benefits. Every school or clinic built, however, will require higher government recurrent spending into the future on the salaries of teachers, medical supplies and so forth. This can be funded from mineral revenue, but there are increased problems if prices fall. The collapse can be dramatic. In 1974, minerals provided over half Zambian government revenue, yet only three years later none at all. This is perhaps not such an immediate problem for Namibia in that since 1978 there has, anyway, been a slump in the mineral economy. Unlike Zambia, Namibia is likely to come to independence with a very clear realisation of how brittle and unstable mineral booms can be.

Policy

The implication of this analysis for general economic policy extends beyond the scope of this booklet. Strong state action to support local production and spread the benefits of development is necessary. Methods used elsewhere include earmarking a proportion of mineral revenue for productive investment as Venezuela did, establishing a stabilisation fund as in Papua New Guinea, and using revenue to subsidise peasant farming. There are, however, two points of direct relevance to mineral policy.

(a) *Diversification* First, a diversified range of minerals gives more reliability. Zambia is almost wholly dependent on copper and cobalt. Namibia already has a range of metals, diamonds and uranium. The 1981/82 crisis shows that even this range cannot prevent a slump in revenues if the world economy itself slumps. But it is significant that the effect on tax revenues would have been less if Rössing had been paying tax — for the price received by Rössing for uranium did not fall at the same time as that of the other minerals.

(b) *New mines or development of other sectors* The second point is that, given Namibia's shortage of skilled and reliable people, there is a definite trade-off between the development of mines and other development projects. Planning, monitoring, and especially building a new mine and support services, are very demanding of senior staff and politicians' time, which could otherwise be devoted elsewhere. There is little point in generating large flows of income if there is not time to use it wisely and if more important priorities are neglected. It is clear that development priorities are likely to be outside mining. For this reason, the new government may not be too disappointed at the lack of immediate prospects for new large mines.[3] It is nevertheless true that the relatively short life remaining to two or three of the largest base metal mines does imply either that a medium term decline in output must be accepted or that proving and development of new mines needs to be begun by the last quarter of the 1980s.

5.2 Contribution of Minerals to Government Revenue

How important are minerals to government revenue? Table A8 shows the position in the last few years. Until the present crisis, mines provided directly over a third, and in 1978/79 as much as 60%, of government income from Namibia; more came indirectly, through loan levy, tax on dividends paid to mining company shareholders outside Namibia, and customs duty on imported mining inputs.[4]

---

3. The standard economist's rule for whether to mine now or later is that minerals should be mined now unless more revenue could be earned by mining later and unless you do not need the money now. The rule is worth remembering — it suggests that mining companies, if prepared to invest at all, will want to mine as fast as possible, fearing that their long term position in Namibia is uncertain. However, the rule should not decide government policy: there are much wider implications of an expansion or decline in mining, which require detailed consideration in each case.
4. The other major contributor in recent years has been customs and excise, paid of course by South Africa. Until 1981/82 Namibia received much less proportionately than paid to Botswana under the Southern African Customs Union Agreement (SACUA) — over 1975 to 1979 annual payments averaged 10% of imports two years earlier (the figure on which SACUA payments are calculated) compared with 24% for Botswana. The 1981 Namibia figure, on the other hand, jumped to 44%.

It is impossible to predict with any accuracy the extra surplus that might be obtained from mines after independence. First, much more information is needed on the present profits of mines: published figures are insufficiently explained, especially for diamonds and uranium, and unobtainable for the South African state-owned mines at Uis and Rosh Pinah. One needs to know, not just 'pre-tax profit', but for example:

(a) What deductions have been made for depreciation (that is, crudely, the extent to which machinery and other capital investment has worn out during the year). To assess whether this is reasonable, one also needs to know capital expenditure over time (when the machinery was bought and how much it cost).
(b) Whether capital expenditure to maintain the mine at its present level of production was or was not deducted before profit was calculated.
(c) What loan repayments and interest payments have been made, and how much would fall due in the future.
(d) Whether the other costs of the mine are reasonable. It is possible to disguise profits as costs — for example, the parent foreign company may charge its Namibian subsidiary exorbitant fees for management services.

If the present profit level of Namibia's mines is not known, even less is known about the future. The prices of minerals and the quantities that can be sold are far from certain: in the past there have been large swings. Finally, it is unknown what arrangements will eventually be reached between the independent government and mining companies or managing agencies.

For all these reasons, a detailed estimate of future mineral revenue would be foolish. Table A4 to A6 in the appendix sets out some past raw data, but they should be interpreted with caution. Nevertheless a few general points can be made.

For *diamonds* much depends on (a) whether the CSO (or another producer cartel) can keep control of the market, (b) how fast demand for gems picks up again, (c) whether new mines elsewhere in the world cause over-supply, and (d) on Oranjemund's level of production, which in 1982 was only 51% of its 1977 level. All being well, reported pre-tax profits could be in R300 to R400m. range. If government increased its share of profits to 75% (as reported in Botswana), diamond revenue would then be between R200 to R300m., depending on what capital expenditure allowances are permitted and the extent to which taxable profits differ from reported pre-tax profits. On the other hand, it is possible to imagine a collapse with revenue R50m. or even lower.[5]

A similar market problem besets *uranium,* where world supply looks likely to exceed demand at least until the early 1990s. With prices maintained, pre-tax profits as currently reported might be between R150 and R200m., perhaps more depending on whether loans have now been paid off. The amount of this which could be taken as state revenue depends again on the extent to which profits for tax purposes prove to differ from the reported figures, but also on the state's attitude — and ownership relation — to Rössing. The present company is likely to argue to a sympathetic government that it requires a high proportion of profits in order to make a reasonable return on its investment, although even a relatively cautious interpretation of its own figures suggests that by the end of 1983 it will have recovered its investment and made at least 15% internal rate of return on equity (at current prices). A radical government might dismiss this argument for a high rate of return on the ground that the investment was illegal. With high tax rates, state revenue would then, in a reasonable world market but with no increase in production, be in the R100 to R150m. region. It is, however, very possible that, in the present world uranium glut, new sales contracts might require prices of say 20% below the 1982 level, or even lower.[6] A 20% price fall in 1982 would apparently have halved reported pre-tax profits; uranium revenue to government could fall below R50m., especially if a rise in the value of the rand increased costs relative to income.

In the recent past, *other minerals* have been a small part of tax revenue. Tsumeb, described in 1975 as having been 'extremely profitable' over the years, was in the five years previously taxed only an average of 31% of its operating income, with no significant increase of the percentage in years of high profit.[7] Since 1975, TCL has paid almost no tax, because of the legal provisions allowing accounting losses in one year to be carried forward for tax purposes, and allowing capital expenditure to be fully deducted before income is taxed. Future revenue from base metals will depend on (a) revision of such provisions, but even more on (b) world economic recovery thus raising metal prices, and (c) the profitability of the mines and their life.[8] A continuation of the present world slump would mean

5. 1982/83 estimates by the present regime are for diamond revenue of R35m. After independence, tax could be increased, but the market position could weaken further.
6. See Chapter 3, footnote 19.
7. From Newmont's submission to the US Securities and Exchange Commission, as reported in Murray 1978.
8. Profitability is large dependent on technical factors: if a mine is low cost, compared with others in the world, then there should be 'rent' to be taxed. If it is a high cost mine, tax will inevitably be limited.

continued low revenues, whatever tax and ownership arrangements are made — possibly in the region of R10m. If the market moved to the opposite extreme, and metal prices double, Tsumeb Corporation Ltd alone would be reporting operating income approaching R100m. With similar performances from Klein Aub, Rosh Pinah, Uis and Oamites, the tax take from base metals could reach R100m.

The major implication of this analysis is the extreme variability of mineral revenue. In a continued world slump, even with new tighter ownership and tax provisions, total mineral revenue could be less than R100m. This compares with likely post-independence expenditure estimated at R1 000 to R1 100m. (in 1981 prices).[9] Given a more prosperous world, however, mineral revenue could reach R500m. — more if it proves that companies have been significantly under-reporting profits. Mineral revenue will be very important to Namibia but dangerously unpredictable.

5.3 The Mineral Sector in an Integrated Economy

The independent government is likely to try to integrate the economy further than at present, if only to reduce Namibia's exposure to the fluctuations and lengthy mineral recessions of the world economy. In pre-colonial times, mining was intricately linked to the rest of production. There will, however, be enormous difficulties in trying to re-integrate the present mines.

1 Inputs

The most promising area is in local production of mining inputs — if only because so little is achieved at present[10]. Mines do buy some of their inputs from traders in Namibia: R40m. (69%) was the figure for Rössing in 1981. But only a tiny proportion (½% or R330 000) of the Rössing total was spent on goods *manufactured* in Namibia: safety slings, overalls, detergents and paints. More is spent on services provided in Namibia, such as repair of hydraulic cylinders (R557 000), rewinding electric motors (R500 000) and catering and cleaning (R1.2m.) Nevertheless, the bulk of requirements come from South

9. Divided equally between recurrent and capital spending, R.H. Green 1981.
10. Collett 1979 estimates that in 1976 the mining sector had a production of R291m. and spent R95m. on purchase of inputs from other sectors, divided as follows:

Manufacturing	*Electricity Gas and Water*	*Construction*	*Trade*	*Transport & Communication*	*Services*
40	8	2	10	19	16

However, some of these sectors, especially manufacturing, clearly had a very high import content themselves. An example of what has grown up is SWE's steel pipe manufacturing capacity (WA 18 Mar. 1983).

Africa or abroad.

CDM explains: 'Expenditure within Namibia would have been higher had it not been for the location of the mine in the south-eastern corner of the Namib Desert, isolated from the country's road and rail network. The transport route from the south is the most economical to use for supplies'; indeed a harbour has been built inside South Africa, at Port Nolloth, through which much of CDM's transport is handled. Excluding wages, only 30% of CDM's R100m. purchases of supplies or services went through Namibian suppliers in 1981: the amount manufactured in Namibia must have been miniscule. Few data are available for other mines, but the lack of industry in the country proclaims the same story.

Even in supplying food for the mines, there are few links to the 'white' commercial agricultural sector, let alone to the small plots and herds on which black farmers depend. Instead, CDM and Tsumeb have their own farms. Rössing does order from a separate firm, quaintly named Alpine Caterers, but its high-technology hydroponic farms are quite different from other Namibian farms. These farms are in effect agricultural enterprises within the mining enclaves. Tsumeb management is quite clear that it does not want to rely on Namibian farmers: its farm is maintained, at an annual loss of R20 per employee, in order to provide a secure food supply for the mine.

Future possibilities

Many mining inputs will continue to be imported, since the sector is too small to support an industry making complex mining equipment. If the main role of the mineral sector is to provide revenue, it serves little purpose to raise its costs by forcing the purchase of expensive locally-produced specialised inputs.

Yet the present lack of integration does offer genuine possibilities, especially if selective tariffs are applied to South African imports. Particular attention might be given to products which have a wider market than mining, in an attempt to diversify the economy as well as reduce unit costs: these might include broad market goods such as building materials, clothing, hard hats, food and goods and services for mine workers. Increased repair facilities, foundries and so forth, are another possible area.

Another possibility is increased inter-mine trade — such as the manganese dioxide from Otjosondu to Rössing, or coal and cement.[11] A third group of products are ones which can be sold to (or imported from) other SADCC states. Chemicals and explosives might be considered.

11. Collett 1979 estimates that R8.5m. of coal was used by power stations in 1976. TCL is investigating cement (FM 22 July 1983).

Active intervention will be required. Satisfactory local facilities for repair of hydraulic cylinders and electric motors were only developed by Namibian firms in the past by direct advice and assistance from Rössing engineers: similar advice and assistance will be needed in future. Contracts must not be too large or complex for small local producers. The state might consider establishing an umbrella organisation to support local production for the mines.

However, diversification of production and increases in national economic integration need to be viewed in an overall political economic framework, not sectorally. If grain, basic consumer goods, construction materials and meat processing prove to have the highest payoff in terms of employment, rural income and reduction of external dependence, they might be expected to have priority over goods required by the mining sector.

2 Local processing of minerals before export

The reason for increasing production of mining inputs is to stimulate a more integrated, less dependent economy as well as to create jobs. The reason for increasing local processing of minerals before export is different. It will in one sense increase dependence on mining, and generally have few links with the rest of the economy. The aim is to obtain more benefit from the minerals themselves by raising earnings per ton mined, and by increasing employment. It often runs against the interests of transnational corporations, who may prefer processing at 'safer' locations, and of industrial countries, who try to retain processing by charging higher customs duties on processed rather than on raw materials. Further, it is not always true that maximising local processing does increase the benefit from minerals: revenue could fall if processing is considerably more expensive in Namibia than elsewhere. Each mineral needs examining separately.

Diamonds are an example where further processing in Namibia might be desirable. Local sorting of the diamonds is a clear case: it would provide a few jobs at a modest cost (perhaps 40 Namibians plus 10 expatriates initially) and above all would enable Namibians to monitor the prices received when they are sold. This is likely to be acceptable to CDM: indeed, the new CDM building in Windhoek has clearly been built so as to be suitable for sorting. The next stage for gem diamonds is cutting or, more properly, 'finishing'. This is much more risky. Finishing would increase the sale value of diamonds. On the other hand, South Africa's cutting industry is said to survive only on subsidy, though it has survived the current recession surprisingly well. Tanzania's small cutting industry ran at a substantial loss and was the most skill-intensive operation in the industrial sector — neither feature a particularly strong recommendation for it.

Nevertheless, the work is labour-intensive, and if an established reliable and independent partner can be found, it could provide a limited amount of employment quite cheaply. In India, it is a cottage industry carried out (for low pay and long hours) by people in their own homes, or in village co-operatives of less than 20 people. Botswana's first cutting factory, established in 1982 with a Belgian investor, employs two expatriates and 63 workers (the figure is due to rise to 200) for a total capital investment of approximately R2m. CDM has argued in the past, and was supported by a 1973 South African Commission of Inquiry, that investment in training sorters and finishers was a mistake, given the short remaining life of the mine. In practice, Botswana's cutters were able to start work after less than a year's training. Likewise, although 10 years' experience is needed for senior sorters, most sorters can begin work under supervision after a few months' training.

Uranium is the polar opposite of diamonds, in that further processing seems technologically almost impossible and geo-politically imprudent. Further processing of Rössing's uranium would be to take a step beyond any other uranium oxide producer. Currently the normal pattern is for mines to sell oxide to the utilities owning nuclear reactors. Thereafter utilities contract out first *conversion* of the oxide into uranium hexafluoride and then *enrichment*. Enrichment is carried out with very complex and secret technology by state-owned concerns in the USA or Western Europe. Even for conversion, the skill and capital requirements and the difficulty of market penetration would far outweigh the rather small benefits. Furthermore, while uranium oxide is neither directly usable in weapons (nor, given proper precautions, very dangerous to handle), more refined forms are, and enrichment might therefore lead to distinctly undesirable complications including rationalisations for raids by South Africa and its allies.

Namibia's *lead* is already refined at Tsumeb, which also smelts the *copper*. A copper refinery in South Africa was proposed to refine the smelted output both of Tsumeb (43 000 tonnes of blister in 1979) and of the closely-linked O'okiep mine (32 000 tonnes) in South Africa. Though the project was abandoned with falling copper prices, Tsumeb management still talk optimistically of a Namibian refinery. However, UNIDO has estimated that the minimum size refinery would require about 60 000 tonnes, so that additional blister would be needed.[12] In present world conditions, the benefits of refining are doubtful. Investment is high, and jobs few. Processing of copper contributes only about 15% of the final selling cost. Existing refineries are under-used, and offer to refine at considerably lower

12. UNIDO 1980.

rates than a new Namibian plant could achieve. The position may change in future. New techniques could take advantage of any cheap electricity Namibia might generate, whilst laws protecting the environment in industrial countries may raise the price of refining elsewhere. The probably best option is (a) to retain the right to review any copper sales agreement should the economics of refining become more favourable, and (b) to ensure that the copper is refined outside Namibia at the lowest possible rate.[13]

Zinc is at present smelted in South Africa. The UNIDO minimum size for a smelter is 30 000 tonnes of metal equivalent, compared with about 22 000 tonnes in Rosh Pinah's concentrate in 1980/81. As with copper, existing smelters in industrialised countries, at present, offer a considerably cheaper rate than a new plant would. *Tin* is also exported unprocessed to ISCOR in South Africa, but at levels far below those needed to render a smelter economic.

3 Mineral processing for local use

As the previous section has indicated, most of Namibia's present mineral output is unlikely to be used locally. In the 1976 Input-Output Table, the only other sector to be supplied by mining was construction, with R1m. worth of quarry materials. Uranium and diamonds are of little direct use to Namibia. It is possible that some of the refined metals could be used in Namibian products: there is, for example, a significant demand for copper wire and cable, and lead has possibilities, albeit with health risks. In the processes with high economies of scale this is unlikely: both Zambia and Chile, for example, have invested in plants in Europe for continuous casting of copper wire rod, rather than set them up at home.

Ever since the 1964 Odendaal Report, observers have pointed to the prospect of a chemical industry based on salt, cheap electricity and preferably coal.[14] This does deserve study, although the particular route of manufacturing soda ash is no longer likely to be sensible since higher oil prices have raised the cost of manufactured soda ash compared to natural soda ash, and Botswana has large deposits of natural soda ash. A regional or other export market would be essential, and the capital cost and the skill requirements would be high. In the short run, government may consider this less of a priority than the more urgent tasks of, for example, education and agriculture.

4 Infrastructure

Mines demand infrastructure — notably transport links, electricity,

---

13. See p. 93 above for a suggestion that this has not been the case in the past.
14. SA 1964. See also Collett 1978.

water and telecommunications. Traffic to and from Tsumeb Corporation plant alone provides no less than 30% of rail traffic in Namibia.

In the past, the major mining related investment — almost the only long term planning in colonial Namibia — has been for water and power. The Department of Water Affairs spent R39m. on capital expenditure in the 1960s, and R194m. in the 1970s. By 1979, 34% of the Department's non-agricultural water went to mines (three-quarters to Rössing) and another 47% to urban areas.[15] The massive Ruacana hydro-electric scheme was built primarily with the intention of supplying electricity for Namibia's industry and mines.[16] Existing plans call for very large investment in two schemes to bring water from the Kunene and Okavango Rivers into central Namibia. Water planning is essential for such an arid country, but so far the beneficiaries have been largely the colonisers; the full promised network of canals and pipelines for agriculture in Ovamboland, for example, has not materialised.[17] Meanwhile, a third of the cost of the Department of Water Affairs water is covered by government subsidy, rather than paid by the consumer, although it is said that in the case of mines the full cost is charged.[18]

The history of these projects — heavy and rising expense with benefits largely limited to mines and towns — carries a warning for the future. Large urban and mining development will require expensive water schemes. Mining companies in particular will try to persuade the state to pay for their infrastructure, arguing that it is not the mines' responsibility and that, anyway, other people will benefit. However, from the nation's point of view, there is usually no point in opening a new mine (or expanding an old one), unless the profits from the mine pay for the costs of its infrastructure. There may in fact be others who gain and this can be estimated and allowed for, but the main

---

15. Directorate of Water Affairs 1979 p. 24. Some mines, including Tsumeb and CDM, have their own water supply, and are not included in these figures. In addition to Rössing (9.5m. m^3 in 1981/82), the Department also supplies Rosh Pinah (1.1m. m^3), Oamites (0.4m. m^3), Uis (0.4m. m^3), Matchless (0.1m. m^3) and Otjihase (0.4m. m^3 in 1981, but 1.5m. m^3 in 1976/77). Since 1979/80 the relative importance of water supply to mines has fallen, partly due to closure of mines, but also due to Rössing cutting water use by 12%. There has also been a significant expansion of water supply to agriculture. Water Affairs, Annual Report 1981/82. Water supply to Uis has been restricted in the past, and an emergency additional supply was being installed because drought meant insufficient water in the Omarusu River. WO, 30 April 1983.
16. But also for export of power to South Africa, which began in 1983 with the completion of a power line from Namibia to South africa. WA 28 Oct. 1982.
17. Moorsom 1982.
18. Directorate of Water Affairs n.d. p.6.

beneficiary is the mine, and it must pay its way.[19]

Infrastructure built for mines is not necessarily of great benefit to the wider community. Infrastructure can simply ignore the rest of the economy and be merely represented by the electricity lines passing above houses. It can even be destructive: the water supply for Rössing comes from the Kuiseb and Omarusu river beds, and although information is restricted, local experts fear that the available water is being over-used, and the underground water near Rössing polluted.

That said, the growth of a mine does offer an opportunity, with careful planning, to spread the benefits of infrastructure to a wider public. The cost of transporting a gallon of water, in particular, decreases quite fast as the size of the pipeline increases. Where it may have been uneconomic to pipe water before (for example for irrigation), the establishment of a mine in the area can lower the unit cost of water to a level where it becomes feasible. Likewise, a new road to a mine may boost other production in the area, by lowering the costs of transporting produce to markets.

Detailed advance studies are required if new mines are proposed, since spin-off benefits may call for additional investment (in feeder roads, or a larger size of water pipe) at the time of construction. In the short to medium term, the most important consideration is likely to be how far the existing and planned water network can be modified to increase reliable water supplies in the northern peasant areas.

Export links

In the immediate aftermath of independence, the most important infrastructure consideration, however, may be the need for port facilities at Walvis Bay. The South Africans claim Walvis Bay and a number of coastal islands as South African territory, but this has been rejected by the United Nations. Walvis Bay is at present the only modern deep-water port in Namibia, and is used for the import of supplies and the export of most metals.[20] If South Africa attempts to keep Walvis Bay and prevent it being used, considerable short-term problems and unnecessary expenditure would be caused. Diamonds and uranium, the chief revenue earners, are anyway sent by air, but alternative arrangements would have to be made for supplies and if possible for metal exports. The old German harbour at Swakopmund could be dredged and re-equipped, and the Angolan port of Mocamedes would be a possibility in the north, but the excision of

---

19. The government of Western Australia, for example, insists that mines provide their own infrastructure.
20. In 1973/74, 181 000 tonnes of ore and minerals were exported through Walvis Bay. Presumably Rosh Pinah and perhaps Uis concentrates travel by rail to South Africa.

Walvis Bay would be bound to cause extensive disruption, as it is an integral part of Namibia and its economy. The new Namibian government would expect the active support of the international community in recovering the port.

Walvis Bay is the logical terminus of a potential major regional project, the Trans-Kalahari Railway.[21] The Government of Botswana is keen to see a line through Gobabis, after Namibian independence, for export of the huge coal reserves in Eastern Botswana, as well as soda ash and several minor mineral deposits. It would provide better internal transport links and direct access to the Atlantic other than through RSA. Botswana commissioned a study of the possibility.[22] Given favourable coal prices and a sufficiently large mine, it is believed the line would be economic, though the capital cost would be extremely heavy. Namibia would then have, for the first time, a major link with the rest of the SADCC region. The benefits would not primarily be to Namibia's own mineral industry. Nevertheless, a supply of cheap coal would be useful, both directly in smelters and indirectly in reducing electricity costs. Namibia's own coal prospect at Aranos might suffer by competition, depending on the grade, but alternatively might become more economic since the Trans-Kalahari railway would pass quite close, and coal export facilities would be available at Walvis Bay. Other advantages of the line to Namibian mines are more speculative: a widening of the market for production of mining inputs; perhaps the sale of minerals eastward within the region (but for some minerals in competition with Zambia and Zimbabwe, with shorter transport routes); and possible processing at Tsumeb of some of the more complex ores from other countries.

Improved transport links with Angola, particularly in roads, are likely to be important to an independent Namibia, especially given the large population close to both sides of the border. With a settlement in Namibia, Angola ought to be a major growth point within SADCC, and trade can be expected to prosper. However, minerals are perhaps unlikely to be as important as other goods in this trade.

---

21. R.H. Green, in Harvey 1981, for a discussion.
22. A second, less detailed survey was undertaken by a private consultancy firm, Loxton Hunting, supposedly on their own account, but in practice in close collaboration with the DTA. Botswana is updating its own more thorough study during 1983/84.

6 Conclusion

As the two last chapters have shown, the mining sector would provide both opportunities and problems for an independent Namibia. Taxes from mines can fund a national development programme to meet the basic needs of Namibians. However, the inherited lack of Namibian control over the mines and the inherited social relationships on the mines both call for large and difficult changes. Managing a mineral-dependent economy, exposed to world booms and slumps, poses further problems; even spreading the benefits of mineral revenues to the population as a whole is not easy. However, there is considerable international experience available, and there can be little doubt that Namibia could look after its golden geese.

None of these possibilities can be realised, however, while Namibia remains under South African occupation. When one reads the reports of prestigious mining companies whose Namibian holdings are discussed alongside those in Britain, Canada or Botswana, or looks at airline timetables with the direct flight between Windhoek and Europe, it is sometimes difficult to remember that Namibia is a country under a military occupation internationally acknowledged to be illegal. A stream of independent reports, from churches and other bodies, confirms the impact of the occupation of Namibia, the suffering of the people, their overwhelming rejection of South Africa and their campaign for liberation.

The role of the international mining companies in this continued occupation was examined in the earlier part of this booklet. There is no doubt that on some mines considerable improvements have been made to workers' wages and living conditions, though the system of contract labour remains the norm and much remains to be done. Yet whatever improvements are made at the mines themselves, it is impossible for the companies to extract themselves from the surrounding society. Mines in Namibia cannot be islands. Companies have to protect themselves by co-operating with the police and armed forces. The presence of the companies confers a spurious respectability upon the South African occupation. The tax payments, especially of the diamond and uranium mines, provided in the past up to half the revenue of the illegal administration — to support bureaucracies whose very civil servants accuse them of profligacy and corruption.[1] The problem, therefore, is not simply the moral one that the authority of the United Nations and the advisory opinion of the International Court of Justice are being ignored. The transnational mining companies are also, whatever their intentions, materially supporting the suppression of Namibian freedom.

Mining might seem very far from the concern of the Christian churches. But the impact of mines on communities and on national life cannot be ignored. The church worldwide has taken a growing interest in major mining projects. The Catholic Bishops' Conference of Panama and Australia, for example, have expressed concern about the wider implications of proposed new mines in their countries, and spoken of the basic needs of people in mining areas.[2]

In 1977 Bishop Richard Wood, the Anglican suffragan bishop of Damaraland, contributed a preface to a report on Rio Tinto-Zinc in Namibia, published in Britain by Christian Concern for Southern Africa.[3] Since then, hopes for a Namibian settlement have risen and fallen. But his conclusion remains:

> There may be truth in the statement that 'Britain has lost an empire and not found a role'. Certainly no creative role is likely to be found unless integrity is the foundation. It would be good to show that we have grown beyond the plunder and slavery of poorer people and are learning to express an international ethic which our past experiences and present need could teach us. Those Western countries at present supporting the economic stability of South Africa share in the responsibility for the continuation of legalised injustice there and in Namibia and in the long run they will lose financially as well as losing the goodwill of Africa. They are already being more commonly spoken of as predators rather than as friends. There is never an easy time to start the process of finding a role. In days of recession and unemployment it is 'impracticable'. In the days of a boom the good returns make it 'untimely'. The time for action is always now.

---

1. A quotation from the *Rand Daily Mail* of 3 June 1983 suggests an absurd extreme: Rössing, who have not until recently been liable for taxes, suddenly, out of the blue, coughed up R2m. for the State treasury. The reason, Eric Lang explained, was that the State treasury went cap in hand to Rössing and rather embarrassedly said that they could not afford to pay month-end salaries of the top-heavy and hopelessly inefficient civil service. He later told me he had gleaned this priceless bit of information from the then Receiver of Revenue himself (currently facing fraud charges for an alleged massive whisky swindle).
2. Bishops Conference of Panama 1981.
3. Jepson 1977.

Appendices

Appendix 1

Geological Environment and the Prospect of New Mines

Namibia is a large country, about two-thirds the size of South Africa itself, and for that reason alone can be expected to have significant mineral potential. The country appears to be relatively favoured, particularly for uranium and so-called base metals including copper, lead, zinc, tin and iron. There is also an extraordinary series of marine diamond deposits, as well as some prospects for economically exploiting coal. However, shortages of water and transport will pose problems for developing mines cheaply, especially in the more remote regions including the north west.

A detailed and up-to-date assessment of mineral potential in Namibia is presumably contained in a Mineral Resources Handbook prepared by the Geological Survey in 1981, but which the South African government has not allowed to be published. A brief, technical description of the geology of the country was produced in 1982, together with a 1980 Geological Map. Discussion here will therefore be very much of a summary.

The sand areas

For the layman, the first point is that much of the north and east of the country is covered by a layer of Kalahari sand and associated calcretes. There is similar sand over much of the southern coastal strip (Map 3). In the extreme north this sand is as much as 450 metres deep; in the east it is typically less than 100 metres. A deep sand and calcrete cover makes mining less likely to be viable, partly because of the cost of removing the sand or sinking shafts, and partly because there are often associated water shortages. Prospecting is also very difficult due to the lack of rock outcrops. An exception is in uranium mining, since uranium can be found in calcretes near the surface. Uranium would hold the most potential for large new mines shortly after independence, and major evaluation has already been carried out on three deposits at Langer Heinrich, Trekkopje and Tubas. Langer Heinrich is planned to to produce about a third of Rössing's present raw ore output, but with a higher grade. Further discoveries are possible. For example, there has been considerable uranium prospecting in Botswana along old river valleys near the Namibian border, suggesting possibilities in north-east Namibia.

Map 3. Geology and Prospects

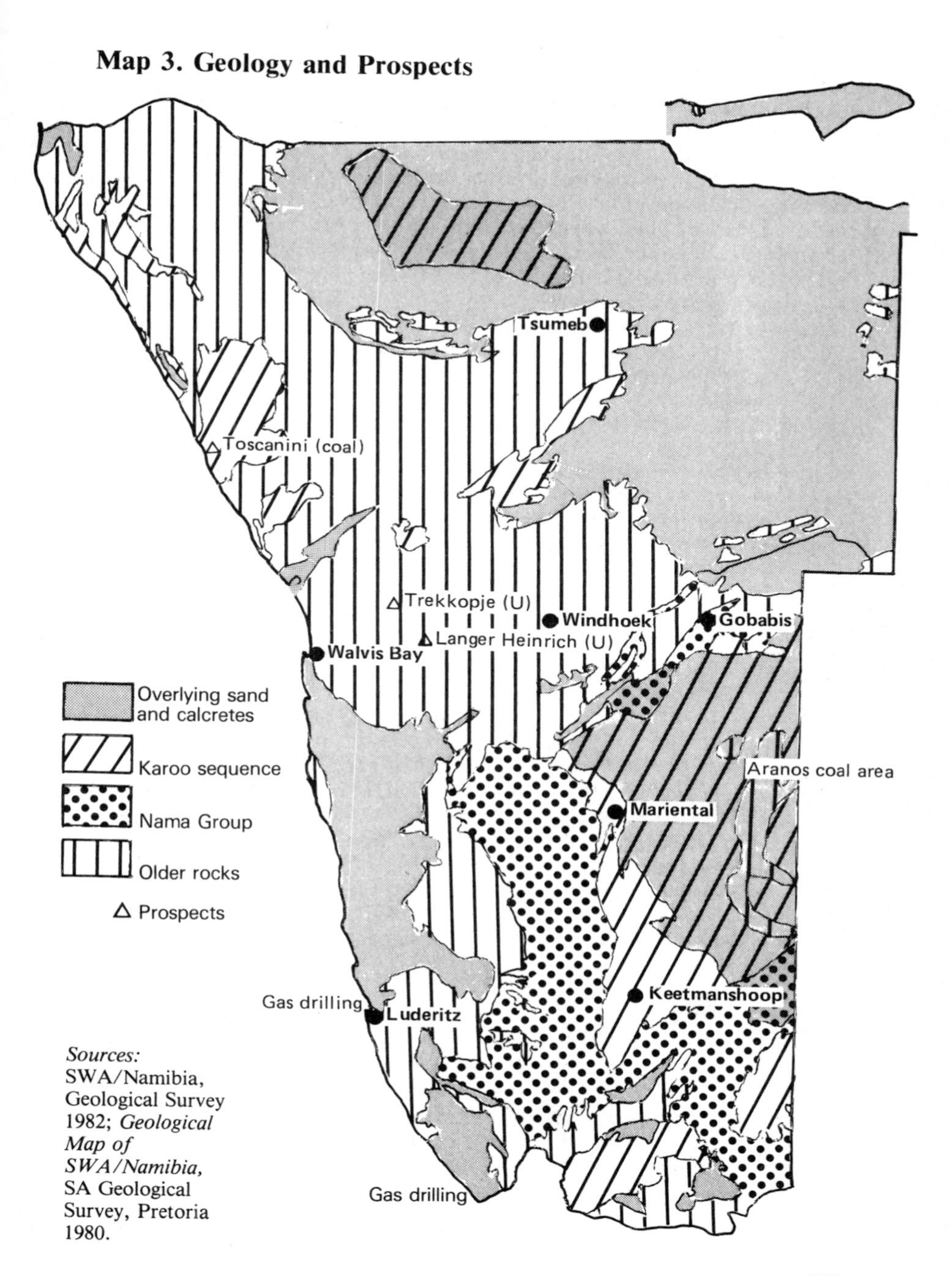

Sources:
SWA/Namibia, Geological Survey 1982; *Geological Map of SWA/Namibia,* SA Geological Survey, Pretoria 1980.

Recent rocks

A second important feature is the geologically young Karoo Sequence and Nama Group of rocks, particularly in the south east where Karoo underlies much of the sand cover. The main mineral potential is coal in suitable parts of the Karoo. The most explored deposits are between Aranos and the Botswana border, where some 25 boreholes have been sunk by CDM, suggesting around 600m. tonnes of good steam (but not coking) coal. At 300 metres depth, economic exploitation would require underground mining and an export market, in order that the mine would be sufficiently large to cover the costs of infrastructure. Ironically, it is also in one of the few areas of Namibia with artesian water, so there would be considerable expense sealing off shafts and tunnels. Coal is also known in Karoo rocks near Toscanini 230km north-west of Swakopmund near the coast, where the Tsumeb Corporation drilled a single borehole and where it holds prospecting rights; in southern Ovamboland, albeit at a considerable depth; and finally, possibly in the eastern Caprivi, as an extension of Zimbabwe's Wankie coalfield. It is not known whether these deposits are economic, because they are either unexplored or the data is not publicly available. It is important that further work be done. The Aranos project, if developed largely for export in conjunction with the Trans-Kalahari railway line from Botswana, would seem likely to be viable by 1990, so long as world coal prices recovered.

If oil and gas are present, they would be in deep basins of similar rocks. Single boreholes have been drilled in the Etosha pan, the most promising area, and off Luderitz, but they were not deep enough. A well off Oranjemund apparently showed wet gas and condensate, and the South African state corporation SOEKOR that was drilling believes that it has established the existence of a gas field.[1] For all three wells complete data has never been made available and what has may not be accurate. The Nossob basin, which Botswana's Geological Survey has identified as having oil potential, also abuts onto Namibia, in the same area as the coal east of Aranos, but the one borehole that has been drilled on the Namibian side was not encouraging. In general, Namibia's prospects of an economic gas field appear fair, but there is no real evidence to justify an optimistic prognosis in respect to petroleum.[1] The UN Centre for Transnational Corporation has recently completed a study on Namibia's oil and gas potential. After independence, co-operation with Angola, which has considerable experience in this field, could also be extremely useful.

Metal potential

The main metal potential can be seen to lie in a wide area from the north-west to the extreme south-east, with an eastward extension through Gobabis. This area contains a variety of older rocks, and the majority of existing mines. Iron and base metals (copper, lead, zinc, tin) are found over a wide area, though, of course, not necessarily in deposits that are economic to mine. Indeed, one problem with Namibian deposits is that most are relatively small, only giving

1. SOEKOR estimates that four step-out wells are needed to establish the viability of the field at a cost of R50m., *Financial Mail* Energy Survey, 13 May 1983.

rise to short-life, medium-size mines, and a need for constant exploration and development to sustain outputs. Manganese has been mined at Otjosondu, tungsten at Brandberg West, and small deposits of gold in Rehoboth. Pegmatites occur, containing tin (Uis), lithium, beryl, columbite-tantalite, rare-earth elements and gemstones (Helikon, Rubikon, Neu Schwaben mines). The Rössing uranium is also in these older rocks.

This whole area, where rock outcrops on the surface, is relatively well-explored. Unlike uranium, it is not thought that there are large, profitable deposits already identified, and which are now ready to be developed at independence. Tsumeb Corporation claims to have spent up to 5% of its budget on exploration since 1947, and yet to have found no new deposits. The mines it has opened were either known before (Kombat, Matchless), or bought from another company who had discovered it (Otjihase).

The future potential of this area is therefore three-fold. First, the discovery of Otjihase only just outside Windhoek demonstrates the possibility of finding new large deposits in the longer term, and the Geological Survey has proposals for prospecting targets. Second, a rise in metal prices could make it profitable to mine known low-grade deposits, which are extensive. Thirdly, rising metal prices might also justify small mines, employing perhaps 50 people each, exploiting known small but high grade deposits.

Diamonds

Namibia's diamond deposits are a geological oddity. They are found in long marine terraces along the sea shore and on the banks of the Orange River — presumably having been washed down the river to the open sea and then deposited on the beach by the northerly-trending sea current. They have now been mined for seventy-five years, and a critical question is how much longer the deposits will last. The mining company (CDM) is extremely secretive, and it will be important for an incoming government to scrutinise their records. Diamonds originate in kimberlite pipes, which come up to the surface of the earth rather like volcanoes. Over 60 kimberlites and similar pipes have been discovered in Namibia, largely around Gibeon, but also in the extreme north east and extreme south east. However, none of them contain diamonds. CDM has conducted a considerable exploration programme for diamonds, but again secrecy prevents a judgment on how thorough it has been. It has regularly proved enough reserves to keep a 15-year life expectancy and in 1982 indicated promising results from prospecting upstream along the Orange River. Government would be wise to review this programme and encourage further exploration, particularly in view of the initial discovery of diamondiferous kimberlites in neighbouring Botswana after De Beers had abandoned prospecting. Semi-precious stones occur in various places and their collection and polishing can provide a limited number of jobs.

Salt and guano

Two very recently formed minerals deserve mention for their practical usefulness. Extensive salt pans occur along the coast, and sodium salts have also been produced for export from Otjivalunda brine pans in Ovamboland. There are guano (bird dropping) deposits on a number of small islands and specially constructed wooden platforms off the coast (incidentally claimed by South Africa, along with Walvis Bay). These are small in absolute terms but are a useful source of fertiliser.

Appendix 2

United Nations Council for Namibia Decree No. 1

For the Protection of the Natural Resources of Namibia

Conscious of its responsibility to protect the natural resources of the people of Namibia and of ensuring that these natural resources are not exploited to the detriment of Namibia, its people or environmental assets, the United National Council for Namibia enacts the following decree:

DECREE

The United Nations Council for Namibia

Recognising that, in the terms of General Assembly resolution 2145 (XXI) of 27 October 1966, the Territory of Namibia (formerly South West Africa) is the direct responsibility of the United Nations.

Accepting that this responsibility includes the obligation to support the right of the people of Namibia to achieve self-government and independence in accordance with General Assembly resolution 1514 (XV) of 14 December 1960,

Reaffirming that the Government of the Republic of South Africa is in illegal possession of the Territory of Namibia,

Furthering the decision of the General Assembly in resolution 1803 (XVII) of 14 December 1962 which declared the right of peoples and nations to permanent sovereignty over their natural wealth and resources,

Noting that the Government of the Republic of South Africa has usurped and interfered with these rights,

Desirous of securing for the people of Namibia adequate protection of the natural wealth and resources of the Territory which is rightfully theirs,

Recalling the advisory opinion of the International Court of Justice of 21 June 1971,[1]

---

1. *Legal Consequences for States of the Continued Presence of South Africa in Namibia (South West Africa) notwithstanding Security Council Resolution 276 (1970), Advisory Opinion, ICJ Reports 1971,* p. 16.

Acting in terms of the powers conferred on it by General Assembly resolution 2248 (S-V) of 19 May 1967 and all other relevant resolutions and decisions regarding Namibia,

Decrees that

1. No person or entity, whether a body corporate or unincorporated, may search for, prospect for, explore for, take, extract, mine, process, refine, use, sell, export, or distribute any natural resource, whether animal or mineral, situated or found to be situated within the territorial limits of Namibia without the consent and permission of the United Nations Council for Namibia or any person authorised to act on its behalf for the purpose of giving such permission or such consent;

2. Any permission, concession or licence for all or any of the purposes specified in paragraph 1 above whensoever granted by any person or entity, including any body purporting to act under the authority of the Government of the Republic of South Africa or the 'Administration of South West Africa' or their predecessors, is null, void and of no force or effect;

3. No animal resource, mineral, or other natural resource produced in or emanating from the Territory of Namibia may be taken from the said Territory by any means whatsoever to any place whatsoever outside the territorial limits of Namibia by any person or body, whether corporate or unincorporated, without the consent and permission of the United Nations Council for Namibia or of any person authorised to act on behalf of the said Council;

4. Any animal, mineral or other natural resource produced in or emanating from the Territory of Namibia which shall be taken from the said Territory without the consent and written authority of the United Nations Council for Namibia or of any person authorised to act on behalf of the said Council may be seized and shall be forfeited to the benefit of the said Council and held in trust by them for the benefit of the people of Namibia;

5.Any vehicle, ship or container found to be carrying animal, mineral or other natural resources produced in or emanating from the Territory of Namibia shall also be subject to seizure and forfeiture by or on behalf of the United Nations Council for Namibia or of any person authorised to act on behalf of the said Council and shall be forfeited to the benefit of the said Council and held in trust by them for the benefit of the people of Namibia;

6. Any person, entity or corporation which contravenes the present decree in respect of Namibia may be held liable in damages by the future Government of an independent Namibia;

7. For the purposes of the preceding paragraphs 1, 2, 3 4 and 5 and in order to give effect to this decree, the United Nations Council for Namibia hereby authorises the United Nations Commissioner for Namibia, in accordance with resolution 2248 (S-V), to take the necessary steps after consultations with the President.

Appendix 3

British Council of Churches Statement on British Mining in Namibia

The following resolution was passed at the BCC Assembly meetings of October 1975 in regard to Namibia and the UK's mining interests in the territory:

The Assembly of the British Council of Churches

Welcomes the statement of British Government policy on Namibia made on 9 June 1975 and in particular the Government's desire 'to secure for the people of Namibia their full, free and independent status':

recalls the decree of the UN Council for Namibia of 27 September 1974 designed to secure for the people of Namibia 'adequate protection of the natural wealth and resources of the territory which is rightfully theirs', and the action of the UN General Assembly of 13 December 1974 in requesting that the member States of the UN should ensure full compliance with the decree;

deplores the fact that the mineral resources of Namibia are being exploited by British firms and their subsidiaries; urges the British Government to undertake a thorough-going examination of the conditions under which mining operations are undertaken by British firms and their subsidiaries, and in particular of the wages and living conditions of their employees, and to publish their report.

Appendix 4

The Grade of Ore Being Mined

This is an example of an apparently purely technical matter in which, in fact, the state has a major interest. One of a mining company's responses to low prices, rising costs, and fears about the future is to mine the best bits first — to concentrate on the higher grade parts of the deposits. This is logical: the more mineral that can be obtained from a ton of ore, the more income there will be to offset the cost of mining that ton. It is possible because the amount of mineral in the rock is different in different parts of the deposit. At a sophisticated mine like Rössing, computers are used to plan which parts to mine, and which to process rather than send to dumps.[1] It seems that from 1981 both CDM and Rössing concentrated on somewhat higher grade areas than before,[2] although there is no way of knowing whether the long term mining plans were altered in response to market conditions. TCL certainly mined its higher grade areas during 1982, with a view to cutting costs.

The disadvantage of picking out high grade areas is that the lower grade deposits left behind may be too expensive, or technically difficult, to mine in the future. This sort of corner-cutting happened at Tsumeb: the Anglo report commented that its high profits between 1965 and 1975 were 'achieved to some extent without regard to future mining problems . . . and now a large proportion of the reserves are in pillars which have to be extracted at relatively high cost.' Under certain conditions, it may in fact be in the national interest to concentrate on high grades. But at other times the mining company may want high profits quickly, even though it would be in the national interest for the mine to last longer.

---

1. 'Because the occurrence of uranium in the (Rössing) ore body is very erratic, the ore is blended by means of adjusting digging rates and shovel positions according to the quality of the ore at specific points. Decisions based on pre-calculated data flows are fed by controllers to the hauler drivers who are directed to a particular shovel for loading, and to the waste stockpile or the primary cannister for disposal.' WA 11 April 1980.

2.

	1979	*1980*	*1981*	*1982*
CDM (carats per tonnes ore)	10.31	9.27	9.95	10.13
Rössing (tons uranium oxide per 1 000 tons treated ore)	0.298	0.291	0.314	0.313

Source: RTZ and De Beers Annual Report

Statistical Supplement

Table A1

Namibia's Mines

Mine	Total Employment 1982[1]	White[1]	Mineral	Principal Ownership
CDM[2]	5 470	1 320	Diamond	De Beers
Tsumeb etc.[3]	6 400	1 410	Copper/Lead/Zinc	GFSA/Newmont
Rössing	3 230	895	Uranium	RTZ/IDC/Total
Oamites	413	52	Copper/Silver	Metorex
Klein Aub	1 057	45	Copper/Silver	Gencor
Uis[1]	(363-436)	(36-109)	Tin	ISCOR
Rosh Pinah[1]	(332-399)	(33-100)	Lead/Zinc	ISCOR
Lithium[1]	(97)	(10)	Lithium	Kloeckner
Tin Tan[1]	(74)	(7)	Tin	?
Salt & Sodalite[1]	(110)	(6)	Salt	Various
Others[1]	(326)	(24)	Various	Various
Total[4]	**c.18 000**	**c.3 900**		
Mines now Closed				
Berg Aukas	724	94	Vanadium/Lead/Zinc	AAC/Goldfields of SA
Krantzberg[1]	(327)	(49)	Tungsten	Bethlehem Steel/Nord Resources
Brandberg West[1]	(300)	(30)	Tin/Tungsten	AAC/Goldfields of SA
Sodalite[1]	(87)	(4)	—	Local
Total	**1 400**	**175**		

Notes

1. Figures in brackets are derived from Chamber of Mines figures for black workers, plus an assumed percentage of whites. For Uis and Rosh Pinah, two alternative assumptions (10% and 25% whites) are used.
2. See footnote 12, Chapter 3.
3. Includes all mines operated by TCL: Tsumeb, Asis Ost, Asis West, Kombat, Matchless and Otjihase.
4. The Chamber of Mines (Annual Report 1982) reported a total of 19 580 people engaged in mining. This figure includes 750 engaged in prospecting, and the remaining discrepancy may perhaps be explained by the Chamber using a higher figure for CDM — see footnote 4.

Sources

Data from companies, Chamber of Mines and ENOK Rehoboth Report 1981.

Table A2

Namibian Mineral Production[1]

Mineral	Unit	1977	1978	1979	1980	1981	1982
		Quantity of Mine Production (mineral content)					
Diamond	000 carats	2 001	1 898	1 653	1 560	1 248	1 014
Uranium oxide	short tons	3 042	3 500	4 980	5 250	5 160	4 910
Copper	000 tonnes	49.2	37.7	41.9	39.2	44.3	49.3
Lead[1]	000 tonnes	41.2	38.6	41.0	47.6	46.5	34.7
Zinc	000 tonnes	38.3	36.6	29.0	25.4	36.4	36.4
Other Minerals[2]							
Tin	000 tonnes	1.0	1.0*	1.0*	1.0*	0.8	0.8
Cadmium	tonnes	88	79	81	70	0	107
Arsenic	000 tonnes	2.6	2.4	2.2	1.2	1.4	1.9
Lithium minerals	tonnes	2 548	n.a.	n.a.	3 000*	n.a.	n.a.
Silver[1]	tonnes	40.3	43.5	49.8	47.0	58.7	n.a.
Tungsten	tonnes	125	117	57	100*	n.a.	n.a.
Vanadium	tonnes	775	462	0	0	0	n.a.

Notes

1. It is difficult to reconcile these figures, from international metal journals, with the various series on production and sales in the TCL Annual Report. In particular, there is no sign in TCL of the drop in lead production — TCL production is given as 41 700 tons in 1981 and 40 600 in 1982. TCL silver sales were 85 tonnes in 1981 and 87 tonnes in 1982.
2. In addition to the minerals here, significant quantities of salt are produced (around 400 000 tonnes in 1980, according to WA 9 Oct. 1981), and other minerals listed on p.29

Sources

World Metal Statistics; World Mineral Statistics; Metallgesellschaft; Company Reports.

*Estimates   n.a. Not available.

Table A3

Value of Mineral Sales

Rand million

Mineral	**1977**	**1978**	**1979**	**1980**	**1981**	**(1982)**
Diamond	354	418	383	399	210	(180)
Uranium	123	136	229	291	279	(413)
Copper	48.2	53.9	69.6	69.8	55.6	n.a.
Lead	20.5	24.3	32.8	32.7	23.4	n.a.
Zinc	11.2	9.5	11.7	8.2	12.6	n.a.
Other minerals[1]	24.4	35.5	48.3	69.8	47.0	n.a.
Total Value						
excluding uranium	459	541	545	579	348	n.a.
including uranium	582	677	774	870	627	n.a.

Note

1. At some stage, salt production at Walvis Bay, equivalent to about R3m. p.a., will have been excluded from the Namibia figures.

Sources

SWA/Namibia 1981 and Loan Prospectus for loans Nos. 11, 12, 13, 18, 19 and 20 for all except uranium and 1982.
Uranium from RTZ Annual Reports, group turnover for Namibia. This is inevitably inaccurate, because the definition of 'turnover' is not necessarily the same as 'mineral sales' and because there are very small RTZ operations in Namibia other than Rössing. Sterling figures are converted to Rand at average rates from South African Reserve Bank quarterly bulletin.
1982 figures from Marriott (diamonds), RTZ Annual Report (Uranium). The Administrator General's 1983-84 budget statement (WO 11 June 1983) said total mineral sales had increased by 14.7%, but it was unclear whether he was referring to the total with or without uranium.

Table A4

CDM: Basic Financial Information

Rand million

	Source	1972	1973	1974	1975	1976	1977	1978	1979	1980	1981	1982
Value of Sales	A, B			122	145	179	354	418	383	399	210	(180)
Profit Before Tax	C	99	154	111								
Tax Estimate[1]	D				38	37	65	147	169	116	108	27
Profit After Tax	C, E	64	97	81			112	162	135	116	35	28
Capital Expenditure	C, F	5	9	25	20		Years 1977-1981: Total R148m.[3]					
Net Assets[2]	C	289	346	373	n.a.	n.a.	n.a.	n.a.	n.a.	n.a.	n.a.	n.a.

Note

1. Includes diamond profit tax and export duty, as well as normal tax minus SWFC rebate (see Table A8). However these are for the financial year starting 1 April, not the calendar year as the rest of the table. Furthermore, these are government advance estimates, not actual payments (see Table A8). The Chamber of Mines 1982 Annual Report gave the 1981 actual figure as R46m. because of the slump (1980 was R115m.). See also Table A14.
2. The replacement value of the whole Oranjemund complex was put at R543m. in CDM 1981.
3. Years 1976-1980 R161m.

Sources:

A. Namibia/SWA Prospectus.
B. Loan Prospectus 1983.
C. CDM Annual Reports. Since 1974 no CDM Report has been published.
D. Government Estimates of Revenue.
E. De Beers Annual Report — Namibian % of group profits.
F. CDM 1982.

Table A5

Rössing[1]: Basic Financial Information

£ Sterling

	1978	1979	1980	1981	1982
Turnover	81	128	161	158	218
Profit Before Tax	4.5	27.2	54	63	91
Tax Paid[2]	0	0	0	0	0
Net Profit to RTZ[3]	2.0	12.6	21.1	21.4	32
Capital Expenditure	Total investment to 1980 said to be US$380m.				
Assets			168	179	229

Notes

1. These figures actually apply to the RTZ Group in Namibia, rather than to Rössing Uranium. Although RTZ's other holdings in Namibia are very small, there appears to be a difference: the 1981 RTZ Annual Report says Rössing's contribution to RTZ's net profit was £20.2m. in 1981 and £21.1m. in 1980 (p.40) — the 1980 figure is the same as RTZ group in Namibia, but the 1981 figure is slightly less. Likewise, the RTZ Annual Report says Rössing's 1981 turnover was marginally higher than 1980, whereas the Table above suggests it was marginally lower.
2. Provision for deferred tax was apparently £9m. in 1980, £17m. in 1981 and £22m. in 1982 if RTZ was still receiving 46½% of net profit.
3. RTZ holds 46½% of shares.

Source: RTZ Annual Reports.

Table A6

TCL — Basic Financial Information

Rand million

	1972	1973	1974	1975	1976	1977	1978	1979	1980	1981	1982
Revenue from Sales	37	59	73	55	45	53	75	110	126	87	106
Operating Income	10.1	21.0	23.8	3.8	4.5	8.2	19.9	29.1	30.2	6.8	3.2
Tax Payments[1]				0.3	−0.1	0.9		12.9	0.2	0	0
Profit after tax, depreciation, exploration, interest charges, exchange loss, etc.	5.0	13.1	14.5	−0.2	0.2	2.2	10.6	21.9	14.1	−3.4	−8.8
Dividends Paid			13.1	0.3	0	0	11.0	16.0	15.5	0	0
Capital Expenditure	1.6	1.3	4.9	10.5	5.7	4.2	3.9	2.8	40.9	14.7	5.0
Shareholders' Equity					27.0	29.2	28.9	34.8	33.4	30.0	41.1

Note

1. Low tax payments result from:
   i. tax being deferred. In 1980, provision for deferred tax was R9.6m. The Annual Report explains 'the provision for deferred taxation results in the main from certain capital expenditures being allowed in full for tax purposes in the year incurred, whereas for accounting purposes they are charged to income by means of annual depreciation provisions.' Depreciation is straight line over 10 years.
   ii. carry forward of accumulated tax losses — at 31 Dec. 1980, this was R10.1m. (R23m. in 1981).

Sources: Tsumeb Corporation Ltd. Annual Reports, Report to Securities and Exchange Commission, 1976, as quoted in Murray 1978.

Table A7

The Growing Importance of Mining in the 1970s
(official South African statistics)

Rand million, 1975 Constant Prices

	1970	**1971**	**1972**	**1973**	**1974**	**1975**	**1976**	**1977**	**1978**	**1979**	**1980**	**1981**	**(1982)**
Total Namibian GDP	560	563	568	553	567	629	647	680	685	693	678	695	(684)
of which:													
Mining and Quarrying[1]	172	160	153	156	151	161	165	223	227	224	215	194	(180)
Mining %	31%	28%	27%	28%	27%	26%	26%	33%	33%	32%	32%	28%	(26%)

Note: 1. This Table is not consistent with Figure 2, which is devised from the earlier *Statistical Economic Review* 1982.
2. 1982 figures derived by applying real growth rates as given in the Administrator General's 1983 Budget speech.

Source: *Statistical/Economic Review,* 1983; WA 9 June 1983.

Table A8

Mining's Estimated Contribution to State Revenue[1] (see also Table A14)

Rand million *(% of own revenue in brackets)*

	75/76	*(%)*	**76/77**	*(%)*	**77/78**	*(%)*	**78/79**	*(%)*	**79/80**	*(%)*	**80/81**	*(%)*	**81/82**	*(%)*	**82/83**	*(%)*	**83/84**	*(%)*
Total Expenditure	171		190		218		320		392		520		818		840		1036	
Total Own Revenue[2]	126	*(100)*	128	*(100)*	179	*(100)*	275	*(100)*	357	*(100)*	316	*(100)*	514	*(100)*	459	*(100)*	539	*(100)*
Diamond Mines[3, 4]	43	*(34)*	43	*(34)*	73	*(41)*	163	*(59)*	188	*(53)*	133	*(42)*	124	*(24)*	33	(7)	41	*(8)*
less payment by government to SWFC[5]	−5		−6		−8		−16		−19		−17		−16		−6		−6	
Other Mines[3]	10	*(8)*	2	*(2)*	1	*(1)*	—	*(—)*	5	*(1)*	15	*(5)*	1	*(—)*	2	*(1)*	30	*(6)*
Customs and Excise	28	*(22)*	36	*(28)*	48	*(27)*	45	*(16)*	49	*(14)*	45	*(14)*	250	*(49)*	250	*(54)*	250	*(46)*

Notes

1. These are estimates, prepared annually in advance for government budgeting purposes. The actual outcome may be different. The 1982 Chamber of Mines Annual Report gives actual payments for 1980/81 as R115m. from diamonds and R17m. from other mines, but the 1981/82 diamond contribution as only R46m and other mines R2m. The 1982 Budget Speech, however, said 1981/82 diamond revenue was only R38m.
2. Excludes loans and 'RSA contribution'. Includes customs and excise.
3. Excludes loan levy (15%) applied in some years (1978/79, 1979/80), non-resident shareholders tax, and indirect effects eg. income tax on workers, tax and customs duty on mining inputs.
4. Includes diamond profits tax, diamond export duty, and head 23.0101 'diamond mines' which is presumably mostly income tax.
5. For historical reasons, the South West Finance Corporation Ltd., wholly owned by CDM, received what is in effect 22% of the diamond profit tax and export duty.

Source

Estimates of Income and Expenditure, presented to National Assembly of South West Africa annually.

Table A9.1

Employment at Rössing Uranium

	Date	**Number of Workers**														
Rough Categories		*Unskilled*		*Part Skilled*			*Semi-Skilled*			*Skilled*			*Professional*			
Grades		**1**	**2**	**3**	**4**	**5**	**6**	**7**	**8**	**9**	**10**	**11**	**12**	**13**	**14-18**	**Total**
Africans	Jan 1983	166	218	172	322	201	372	125	51	6	5	0	2	1	2	**1643**
	Jan 1980	282	298	198	306	203	265	70	9	4	1	1	0	0	0	**1637**
Coloureds	Jan 1983	5	13	9	48	66	102	117	104	36	63	10	2	1	1	**577**
	Jan 1980	7	28	50	80	84	138	145	36	29	22	1	1	0	0	**621**
Whites	Jan 1983	0	1	13	10	18	24	48	59	33	233	148	107	47	98	**839**
	Jan 1980	0	5	7	4	22	34	32	83	47	317	133	86	57	59	**886**
Total	Jan 1983	171	232	194	380	285	498	290	214	75	301	158	111	49	101	**3230**
	Jan 1980	289	331	255	390	309	437	247	128	80	340	135	87	57	59	**3144**

Sources: Rio Tinto-Zinc, Fact Sheets No. 2, 1980 and 1983.

Table A9.2

Distribution of Employees and Changes at Rössing January 1980-January 1983

	Unskilled		*Part-skilled*		*Semi-skilled*				*Skilled*			*Professional*		
Grade	**1**	**2**	**3**	**4**	**5**	**6**	**7**	**8**	**9**	**10**	**11**	**12**	**13**	**14-18**
January 1983 Distribution of Employees (% within the grade)														
Africans	97	94	89	85	71	75	43	24	8	2	0	2	2	2
Coloureds	3	5½	5	13	23	20	40	49	48	21	6	2	2	1
Whites	0	½	7	3	6	5	17	28	44	77	94	96	96	97
Change in Proportion — 1980-83 (Increase or fall of % points within each grade — e.g. proportion of Africans in grade 1 fell by 1, from 98% to 97%)														
Africans	−1	+4	+11	+6	+5	+14	+15	+17	+3	+1	−1	+2	+2	+2
Coloureds	+1	−3	−15	−8	−4	−11	−18	+20	+12	+14	+6	+1	+2	+1
Whites	0	−1	+4	+2	−1	−3	+3	−37	−15	−16	−5	−2	−4	−3

Source: Table A9.1

Table A9.3

Engagements and Promotions at Rössing[a]

	Year	Unskilled		Semi-skilled		Skilled	First Professional	Total
Grades[a]		2	3-4	5-6	7-8	9-11	12-13	2-13
No. of Promotions	1981	68	194	345	68	89	29	793
	1980	54	145	162	51	87	26	525
Promotions as % of No.of Employees in Grade promoted into	1981	23	32	42	14	17	18	27
	1980	19	24	22	13	17	17	19
Promotions as % of No. of New Entrants[b]	1981	61	56	76	67	42	69	63
	1980	67	57	56	64	41	53	54

Notes

a. Excludes grades higher than 13.

b. The other group of new entrants are those engaged from outside.

Sources: Rössing Fact Sheet No.2 (1981) and Rössing graphs of engagements and promotions.

Table A9.4

Rössing: Types of Skill Required

Category	Grade	% Whites 1983	% Women 1982	Examples of Posts
Unskilled	1	0	5	Labourer
	2	1	3	Labourer
Part-skilled	3	7	2	Rotoscoop mechanic level 1 (L1); Rubberliner (1)
	4	3	4	Operator; Welder (L1); Electrician (L1); Storeman (L1); Rubberliner (L2); Slurry truck driver.
Semi-skilled	5	6	13	Rotoscoop Mechanic (L2); Welder, Electrician (L2); Driver of Front End Loader, Grader; Rock jack; Barmaid.
	6	5	6	Rubberliner (L3); Welder (L3); Storeman (L2); Driver of Euclid 85.
	7	17	13	Rotoscoop mechanic (L3); Electrician (L3); Boilermaker (L1); Shovel Senior Operator; Instructor.
	8	28	11	Senior Rubberliner (L4); Electrician (L4); Assistant Storekeeper (L3); Boilermaker (L2); Truck Despatcher.
Skilled	9	44	35	Artisan Welder; Overseer Miner; Assistant Foreman, rubberlining.
	10	77	4	Artisan Electrician; Fitter; Boilermaker; Storekeeper; Assistant Foreman, open pit; Translator.
	11	94	2	Charge Hand; Specialist; Computer Programmer.
	12	96	5	Geologist; Junior Engineer; Senior Training Officer.
	13	96	2	Professional and Management.
	14-18	97	n.a.	Professional and Management.

Table A10

Tsumeb: Approximate Labour Breakdown and Training 1982

Category	Absolute No.	% of Total	% White	Minimum Years Schooling for Entry	Training Courses	Length (Years)	Output 1980	Output 1981	No. of Training Officers
Graduate Staff	500	7	100	12	Bursary Scheme				Outside
					Technicians training	3-4	7	8[a]	mine
					Correspondence courses				
Specialists	1 000	15	87	10	Apprentice School	3½	20	5[b]	6
					Upgrading programme	.4		(20)[d]	
					Mining School		10	7	6
Aides	800	12	0	6	Trade Skill School	n.a.	60	0[c]	
					Mining Skill School	n.a.	120	20[c]	3
					Driving School	n.a.	170	110[c]	
Labourers	4 500	66	0	0	Familiarisation and Aptitude Testing	—	(800)	(1 170)[e]	4
Total	6 800	100	22						19

Notes

a. In 1982 25 students studying at university, 6 at Technikon.
b. Low figure due to Government delay in testing, to be made up in 1982. Approx. 40-50% of apprentices are black.
c. Low figure due to 'saturation' — i.e., no posts available.
d. Expected *future* output from new scheme, upgrading aides. Approx. 70% black.
e. Increase due to Otjihase re-opening.

Source: Tsumeb Corporation.

Table A11

Comparative Size of Training Programmes

Turnover and Annual Output of Trainees as % of Employment in Their Grades

	Tsumeb 1981		Rössing 1981		Zambia 1978
	Output p.a. %	Turnover p.a. %	Output p.a. %	Turnover p.a. %	Output p.a. %
Graduates etc.	2	27	3[c]	20[n]	5.9
Skilled Apprentices			4.8[d]	26[m]	
Total Specialists	6[a]				6.8[f]

Total No. of Trainees at One Time as % of Total Mine Employment on diamond mines 1982

	CDM	Botswana Orapa
Graduates & Diploma	0.3[g]	4.2
Apprentices	1.4[h]	2.6[k]

Notes

a. Actual figure in 1981 was lower, because of testing delay.

c. Total cadets 1982 (33) divided by 4 for annual output, as % of grades above 12.

d. 48 apprentices in training (1981) divided by 3½ for p.a. output, plus c. 12 p.a. upgraded through Operator programme, as % of grades 9-11.

e. 101 graduates joined industry after training, (CISB), as % of 1 700 university graduates in mining. (Third National Development Plan).

f. 237 artisans and 153 technicians and technologists joining (CSIB), as % of 3 600 artisans or semi-skilled plus 2 100 technologists or technicians (TNDP).

g. 18 students on degree or diploma studies, 0.5% if 12 accounts trainees included.

h. 77 apprentices. Also a Mechanic Training Scheme to train semi-skilled handy-men, with 250 workers receiving instruction at 5 grades. CDM also set up the Valambola Training Centre in Ovamboland to do artisan training. It is administered by the Ovambo administration, and is not training specifically for CDM. There are facilities for 250 students, but only 66 enrolled for 1983, and the Institute was concentrating on theoretical and academic training.

k. 99 apprentices. Intake now increased to 60 p.a.

m. Turnover 1981 grades 9-11.

n. Turnover 1981 grades 12-13.

Sources: Tsumeb; CDM; Rössing; Zambian Third National Development Plan; CISB Annual Reports; Debswana Annual Report; WA.

Table A12

Training Programmes

CDM 1981/82

64 full-time staff
R1.9m. budget in 1981.

18 degree or diploma students in engineering and management.

77 apprentice diesel mechanics, electricians, carpenters, bricklayers, plumbers, boilermakers, vulcanisers and welders.

In 1980, 250 semi-skilled handymen.

8 trainee instrumentation technicians; 12 accounting trainees.

Specialist in-service training in technical, management, supervisory and safety fields.

Driver and operator training on all vehicles — 1981 over 2 000 workers on initial or refresher courses.

General adult education programmes, including mathematics.

Rössing 1981

56 full-time staff.

33 degree or diploma students.

70 apprentices in six trades.

Semi-skilled training for 1 492 employees 1979-82, of which 1 262 have completed.

Specialist in-service training in technical, management, supervisory and safety.

105 workers on literacy courses 1983.

96 employees having correspondence courses paid for.

Tsumeb 1981

About 19 staff.

For more detail, see Table A10.

Table A13

Who Benefits from Mining (based on official South African data)

	Average 1974-78 (%)	1970	1971	1972	1973	1974	1975	1976	1977	1978	1979	1980	1981
							Rand million						
Value of Sales	(100%)	120	106	143	244	212	225	302	582	677	774	870	627
— implied input costs	(32%)												
Gross Value Added		105	86	117	187	164	163	205	349	531	579	633	409[1]
— depreciation[3]	(4%)	5	5	6	7	9	12	17	21	25	29	32	39[1]
— net operating surplus (before tax after depreciation)	(47%)	76	54	98	149	112	98	115	235	411	444	468	215[1]
— payment of workers[4]	(17%)	24	26	29	32	43	53	73	93	94	106	133	155[1]
Value of Fixed Capital (year end)[3]	—	73	74	79	89	137	243	323	378	440	464	544	570[2]

Notes
1. Assumes 4th quarter bears same relation to the first three-quarters as in 1980.
2. End of 3rd quarter.
3. The generous capital write-off provisions may mean depreciation is overstated and value of capital understated.
4. Of this, only about a quarter was going to black mineworkers in 1977.

Sources
Value of sales from Murray, 1978 (1974 and 1975), Leistner (1970, 1972), Thomas (1971, 1973) and Table A3. All other figures from *Statistical/Economic Bulletin 1982.*

Table A14

Mining's Actual Contribution to State Revenue[1]

Rand million

	75/76	76/77	77/78	78/79	79/80	80/81	81/82	82/83
Total Own Revenue[2]	139	170	217	331	338	292	436	449
Diamond Mines[3, 4]	45	63	108	199	175	135	53	46
less payment by government to SWFC[5]	−6	−7	−15	−19	−19	−16	−7	−5
Other Mines[3]	7	0.3	0.1	1	8	17	2	2
Customs and Excise	30	36	46	47	45	42	258	250

Notes

1. These are actual payments. Compare with Table A8.
2. Excludes loans and 'RSA contribution'. Includes customs and excise.
3. Excludes loan levy (14%) applied in some years (1978/79, 1979/80), non-resident shareholders tax, and indirect effects eg. income tax on workers, tax and customs duty on mining inputs.
4. Includes tax on income and diamond export duty. It is not clear whether diamond profits tax is included.
5. For historical reasons, the South West Finance Corporation Ltd., wholly owned by CDM, received what is in effect a 22% refund of the diamond profit tax and export duty.

Source: *Statistical/Economic Review* 1983.

Bibliography

Books and articles

American Committee on Africa, *Papers* from International Seminar on the Role of the Transnational Corporations in Namibia (Washington, 1983)

Barrera, M., *'Worker Participation in Company Management in Chile: An Historical Experience',* United Nations Research Institute for Social Development, Participation Programme Occasional Paper (Geneva, 1981)

Basson, R. and Varon, B., *The Mining Industry and the Developing Countries,* IBRD (Washington, 1977)

Blom, A.J. and Knights, A.J., *Long term manpower forecasting in the Zambian mining industry,* NCCM Operations and Research Department (Lusaka, 1975)

Brown, R. and Faber, M., 'Changing the Rules of the Game: Political Risk, Instability and Fair Play in Mineral Concession Contracts', *Third World Quarterly* II, 1, 1980.

Brown, R. and Faber, M., *Some Policy and Legal Issues Affecting Mining Legislation and Agreements in African Commonwealth Countries,* Commonwealth Secretariat (London, 1977)

CANUC, *The Rossing File,* by Alun Roberts (London, 1980)

CDM, *CDM in Namibia,* (Windhoek, 1981)

CDM, *Faith in the Future,* (Windhoek, 1982)

Christie, R., *Research Report, Tsumeb Corporation Ltd Namibia,* mimeo (Oxford, n.d.)

CIIR/BCC (Catholic Institute for International Relations/British Council of Churches), *Namibia in the 1980s* (London, 1981).

Collett, S., *The Economy of South West Africa: Current Conditions and Some Future Prospects,* typescript (Johannesburg, 1978)

Collet, S., *Input-Output Table 1976,* mimeo (Johannesburg, 1979)

Cronje, G. and S., *The Workers of Namibia,* International Defence and Aid Fund (London, 1979)

Crowson, P., *Non-fuel Minerals and Foreign Policy,* Royal Institute for International Affairs (London, 1977)

Daniel, P., *Africanisation, Nationalisation and Inequality* (London, 1979)
Daniel, P., 'Mining and Mutual Interests', *Lome Briefing 8,* EEC/NGO Liaison Committee (London, 1983)
Epstein, E.J., *The Diamond Invention* (London, 1982)
Gordon, R.J., *Mines, Masters and Migrants* (Johannesburg, 1977)
Gottschalk, K., 'South African Labour Policy in Namibia 1915-1975', *South African Labour Bulletin,* Vol.4, 1&2, 1978.
Green, D., *Practical Experience in Zambia in Geological Survey,* Paper for UNIN Seminar on Mining Industry of Namibia (Lusaka, 1983)
Green, R.H., 'Economic Co-ordination, Liberation and Development: Botswana-Namibia Perspectives', in C. Harvey (ed), *Papers on the Economy of Botswana* (London, 1981)
Green, R.H., *From Südwestafrika to Namibia,* SIAS Research Report No.58 (Uppsala, 1981)
Green, R.H., *Namibia, A Political Economic Survey,* Institute for Development Studies, Sussex, Discussion Paper 144 (Brighton, 1979)
Green, R.H., Kiljunen, M.L., and Kiljunen, K., *Namibia: The Last Colony* (London, 1981)
Green, R.H. and de la Paix, J., 'A Nation in Agony: The Namibian People's Struggle for Solidarity, Freedom and Justice', *Development Dialogue* 1982 (1)
Green, T., *The World of Diamonds* (London, 1981)
Jepson, T.B., *Rio Tinto-Zinc in Namibia,* Christian Concern for Southern Africa (London, 1977)
Jones, C.R., *The Botswana Geological Survey Department Throughout the 1970s,* Botswana Geological Survey (Lobatse, 1979)
Jornal do Angola, account of Angolan mineral policy (Nov/Dec 1981)
Kane-Berman, J., *Contract Labour in South West Africa,* South African Institute of Race Relations (Johannesburg, 1972)
Krogh, D.C., 'The National Income and Expenditure of SWA', *South African Journal of Economics,* Vol.28, 1960
Leistner, G.M.E., 'SWA/Namibia's economic problems viewed in Africa context', *Bulletin of the Africa Institute of South Africa,* Vol.21, 11 and 12, 1981
Lewis, S.L., 'Mineral Development, Mining Policies and Multinational Corporations: Some Thoughts for Zimbabwe', *Research Memorandum No.75,* Center for Development Economics, Williams College (Williamstown, 1980)
Lipton, C.J., *Monitoring the Performance of Mining Agreements,* UNTC/Commonwealth Secretariat Workshop on Mining Legislation and Mineral Resources Agreements, typescript (Gaborone, 1978)
Mikesell, R.F., *New Patterns of World Mineral Development,* British North America Committee (London, 1979)
Moorsom, R.J.B., 'Namibia in the Front Line', *Review of African Political Economy* 17, 1980
Moorsom, R.J.B., 'Underdevelopment and class formation: the birth of the contract labour system in Namibia, 1900-26', *York University, Centre for Southern African Studies, Collected Seminar Papers,* 5, 1978/9

Moorsom, R.J.B., 'Underdevelopment, contract labour and worker consciousness in Namibia 1915-72', *Journal of Southern African Studies,* 4(1), 1977
Moorsom, R.J.B., *Transforming a Wasted Land (A Future for Namibia, 2: Agriculture),* London, 1982
Mosha, F.G.N., *The Possible Role of UNCTC in the Mining Industry of Independent Namibia,* Paper for UNIN Seminar on Mining Industry of Namibia (Lusaka, 1983)
Murray, Robin, *Multinationals Beyond the Market* (London, 1981)
Murray, Roger, *The Mineral Industry of Namibia: Perspectives for Independence,* Commonwealth Secretariat (London,1978; revised edition forthcoming)
National Council of Churches, *Tsumeb: A profile of US contribution to underdevelopment in Namibia,* (New York, 1973)
Nicaragua Solidarity Campaign, *The Open Veins of Nicaragua* (London, 1980)
OECD, *Uranium Resources, Production and Demand* (Paris, 1982)
Panama Bishops' Conference, *A Choice for Panama,* CIIR Church in the World Series 9 (London 1981)
PROSWA Namibia Foundation, *Mining in SWA/Namibia* (Windhoek, 1978)
Radetzki, M., 'Has Political Risk Scared Mineral Investment Away from Deposits in Developing Countries?', *World Development,* Jan. 1982.
Radetzki, M., *Uranium: A Strategic Source of Energy* (London, 1981)
Rössing Foundation, *Proposals for Rural Development Programmes: Ovamboland, Kavango, Namaland and Damaraland,* by Miriam Ferreira, 1982.
Rössing Uranium Ltd., *An Introduction to Rössing* (Windhoek, 1980)
SACBC (Southern African Catholic Bishops' Conference), *Report on Namibia* (Pretoria, 1982)
Seers, D., *The Life Cycle of a Petroleum Economy,* Discussion Paper 139, Institute of Development Studies, University of Sussex, 1978.
Sohnge, G., *Tsumeb: An Historical Sketch* (Windhoek, 1967).
South African Labour Bulletin, *Focus on Namibia,* Vol.4 (1 and 2), 1978
Strom, Tor, *Newmont Mining Corporation: Southern African Operations,* dissertation, Columbia Business School, 1972
SWAPO of Namibia, *Constitution* and *Political Programme* (Lusaka, 1976)
SWAPO of Namibia and Namibia Support Committee, *Trade Union Action on Namibian Uranium* (London, 1981)
SWAPO of Namibia, Dept of Information and Publicity, *To Be Born A Nation: the Liberation Struggle for Namibia* (London, 1981)
Thomas, W.H., *Economic Development in Namibia* (Munich, 1978).
Tsumeb Corporation Ltd., *Tsumeb, Copper Town of SWA/Namibia* (Windhoek, n.d.)
USBM (United States Dept of the Interior, Bureau of Mines) *Mineral Facts and Problems* (Washington)
USBM, 'The Mineral Industry of Namibia', *Mineral Year Book Vol.III, Area Reports International* (Washington, 1980)
Wood, B., *International Capital and the Crisis in Namibia's Mining Industry,* in American Committee on Africa, 1983 (above)

Newspapers, journals and annual reports

FM: Financial Mail, Johannesburg
FT: Financial Times, London
MAR: Mining Annual Review, London
RDM: Rand Daily Mail, Johnnesburg
WA: Windhoek Advertiser
WO: Windhoek Observer

Metallstatistik, Metallgesellschaft AG, Frankfurt
Namib Times, Walvis Bay
World Metal Statistics (London)
World Mineral Statistics, Institute of Geological Sciences, London

Chamber of Mines of SWA/Namibia, *Annual Report* (Windhoek)
Consolidated Diamond Mines of South West Africa Ltd., *Annual Report* (last one 1974, thereafter incorporated with De Beers) (Kimberley)
De Beers Consolidated Mines Ltd., *Annual Report* (London)
Newmont Mining Corporation, *Annual Report* (New York)
Rio Tinto-Zinc PLC, *Annual Report* (London)
Rio Tinto-Zinc PLC, *Fact Sheet No.2: Some Aspects of Rössing Uranium Ltd,* issued annually at AGM (London)
Rössing Uranium Ltd., *Rössing,* three times p.a. (Windhoek)
Tsumeb Corporation Ltd., *Annual Report* (New York)

United Nations Publications

Aulakh, H.S. and Asombang, W.W. *Mineral Development Strategy Options for an Independent Namibia,* Working Paper, United Nations Institute for Namibia (UNIN), (Lusaka, 1983)
ILO, *Labour and Discrimination in Namibia,* (Geneva, 1977)
Savosnick, K., *Economics of the Namibian Diamond Industry,* UNCTAD (Geneva, 1978)
UNIDO, *Mineral Processing in Developing Countries* (Vienna, 1980)
UNIN, *Manpower Estimates and Development Implications for Namibia,* Based on the work of R.H. Green, N.K. Duggal (ed.) (Lusaka, 1978)
United Nations Council for Namibia, *Report on the Panel for Hearings on Namibian Uranium,* United Nations General Assembly A/AC.131/L.163 (Part 2) New York, 1980
Ushewokunze, C.M., revised by M.D. Bomani and M.F. Sichilongo, *Draft Report on Legal Aspects of Namibia's Mining Industry,* UNIN (Lusaka, 1981)
Zorn, S., *The Mineral Sector in Namibia and Strategic Options for an Independent Government,* UN Office of Technical Co-operation (New York, 1978)

South African official publications

(a) Annual reports

SA, Bantu Mining Corporation (and later titles), *Annual Report*

SWA Administration, Department of Water Affairs, *Annual Report/Jaarverslag*

SWA Admin, Central Revenue Fund, *Estimates of Revenue and Expenditure for Financial Year Beginning 31 March* (for earlier years, same title but issued by SA Dept of Finance, SWA Account)

SWA Administrator, Report to League of Nations on Mandate (various titles, 1917-47)

SWA/Namibia, Dept of Finance, *Statistical/Economic Review* (1981, irregular)

(b) Other publications

Administrator-General of SWA/Namibia, Volkskas Merchant Bank, Union Acceptances Ltd., *Prospectus for Private Placing of Local Registered Stock Loans No.18, 19 and 20* (Johannesburg, 1983)

ENOK, First National Development Corporation, *'N Streekstudie van die Rehoboth Gebiet . . .* (Windhoek, 1981)

ENOK, *Investment in SWA/Namibia* (Windhoek, 1982)

SA, *Commission of Inquiry into the Diamond Industry of the Republic of South Africa and Territory of SWA,* RP84/1973

SA, Odendaal *Commission of Inquiry into SWA Affairs* (RP12-64, 1964)

SA, Mines, Works and Minerals Ordinance 1968 (Ordinance 20 of 1968) of SWA (as amended)

SA, Railways and Harbours Administration, *SWA — Its Attractions and Possibilities,* (Johannesburg, 1937)

SWA/Namibia, Directorate of Water Affairs, *25 Years of Water Supply to South West Africa 1954-1979,* (Windhoek, n.d.)

SWA/Namibia, Dept of Finance, *On the Economic Front* (Windhoek, irregular, Nos. 1 to 5 issued)

SWA/Namibia, Geological Survey, *The Geology of South West Africa Namibia* (Windhoek, 1982)

SWA/Namibia, Information Service, *Mining Brief* (Windhoek, 1980)

Wagner, P.A., *Geology and Mineral Industry of SWA,* SA Geological Survey Memoir No.7, 1916

Index